# Lecture Notes in Mechanical Engineering

**Lecture Notes in Mechanical Engineering (LNME)** publishes the latest developments in Mechanical Engineering—quickly, informally and with high quality. Original research reported in proceedings and post-proceedings represents the core of LNME. Volumes published in LNME embrace all aspects, subfields and new challenges of mechanical engineering. Topics in the series include:

- Engineering Design
- Machinery and Machine Elements
- Mechanical Structures and Stress Analysis
- Automotive Engineering
- Engine Technology
- Aerospace Technology and Astronautics
- Nanotechnology and Microengineering
- Control, Robotics, Mechatronics
- MEMS
- Theoretical and Applied Mechanics
- Dynamical Systems, Control
- Fluid Mechanics
- Engineering Thermodynamics, Heat and Mass Transfer
- Manufacturing
- Precision Engineering, Instrumentation, Measurement
- Materials Engineering
- Tribology and Surface Technology

To submit a proposal or request further information, please contact the Springer Editor of your location:

**China:** Ms. Ella Zhang at ella.zhang@springer.com
**India:** Priya Vyas at priya.vyas@springer.com
**Rest of Asia, Australia, New Zealand:** Swati Meherishi at swati.meherishi@springer.com
**All other countries:** Dr. Leontina Di Cecco at Leontina.dicecco@springer.com

To submit a proposal for a monograph, please check our Springer Tracts in Mechanical Engineering at http://www.springer.com/series/11693 or contact Leontina.dicecco@springer.com

**Indexed by SCOPUS. All books published in the series are submitted for consideration in Web of Science.**

More information about this series at http://www.springer.com/series/11236

Ritunesh Kumar · Adarsh Kumar Pandey ·
Ravi Kumar Sharma · Gavendra Norkey
Editors

# Recent Trends in Thermal Engineering

## Select Proceedings of ICAST 2020

Springer

*Editors*
Ritunesh Kumar
Department of Mechanical Engineering
Indian Institute of Technology Indore
Indore, Madhya Pradesh, India

Adarsh Kumar Pandey
Research Centre for Nano-Materials
and Energy Technology
Sunway University
Subang Jaya, Selangor, Malaysia

Ravi Kumar Sharma
Department of Mechanical Engineering
Manipal University
Jaipur, Rajasthan, India

Gavendra Norkey
School of Mechanical Engineering
Lovely Professional University
Phagwara, Punjab, India

ISSN 2195-4356 ISSN 2195-4364 (electronic)
Lecture Notes in Mechanical Engineering
ISBN 978-981-16-3131-3 ISBN 978-981-16-3132-0 (eBook)
https://doi.org/10.1007/978-981-16-3132-0



This Springer imprint is published by the registered company Springer Nature Singapore Pte Ltd.
The registered company address is: 152 Beach Road, #21-01/04 Gateway East, Singapore 189721, Singapore

# Preface

This book is aimed to facilitate the scientists, researchers, academicians, industrialists and students to share their advancements in the field of mechanical engineering from all around the world and challenges in sustainable technological developments. This book also provides a new approach in the sustainable direction, i.e., a review analysis of the emerging technologies in the energy sectors as well as mathematical and experimental work. This book also emphasizes the processes that makes the energy-efficient automobiles, waste management and recycling practices.

The first chapter of this book gives an insight into the application of waste rice straw to prepare the thermally insulated blocks for the construction purpose having a very low thermal conductivity. It presents an experimental work of the preparation and characterization of the thermal properties of blocks prepared by the waste rice straw. The next two chapters (second and third) deal with the advancements in the IC engines, where an experimental study on the use of hydrogen enrichment on the performance and emissions of CI engine is presented in the former, while a numerical study on the EGR module to run the thermoelectric generator is reported in the third chapter. The performance of the liquid desiccant dehumidification system integrated with the cellulose pads is discussed experimentally in the fourth chapter, where an enhancement in the coefficient of performance and moisture removal rate is achieved for the sustainable system. Wind turbines contribute a very large portion of the renewable energy worldwide, and a quest of enhancing their efficiency always motivates the researchers. The fifth chapter deals with the same quest and presents a numerical analysis of the modified blade geometry for the vertical axis wind turbine. For the last two decades, nanoparticles are considered as the thrust area in all types of thermo-fluid systems due to their abilities to achieve the required thermo-physical properties and promising stability characteristics. A similar trend is followed in the sixth chapter which presents the numerical work on the use of nanoparticles in the natural convection heat transfer and presents very interesting results about the mechanism of heat transfer with the nanoparticles.

To reduce the dependency on the conventional fuel, the seventh chapter discusses the employment of bio-fuels such as algae as a blend in the diesel fuel and presents the combustion and emission performance of CI engine through an experimental approach. The eighth and ninth chapters focus on the application of solar energy using the experiments, where the former study used the potential of solar energy in a dryer, while the latter used the same for the distillation of water using the copper fins in a pyramid solar still. An efficient design of buildings can reduce the energy consumption as well as can also provide the strength to the building against the winds of higher velocities. Similar work is presented in the tenth chapter, which reported a simulation work on the interference effect of wind pressure distribution. Sustainability not only is about the new efficient innovations to generate the energy or energy conversion devices, but also deals with the efficiency improvement of the existing machinery, like the eleventh chapter presents a simulation work to reduce the vibrations of vertical washing machine using a PID controller. The twelfth chapter reports the numerical analysis of downdraft gasifier to investigate the flow behavior of gases. Northern part of India is categorized as the hot and humid climate conditions which required the dehumidification system to maintain the thermal comfort conditions for human. Therefore, the thirteenth and fourteenth chapters present the experimental work on the hybrid desiccant system based on the vapor compression refrigeration cycle for the cooling and dehumidification of air. Another work on the wind turbine is presented in the fifteenth chapter where a bladeless turbine is studied for the power generation using the numerical simulation. The sixteenth and seventeenth chapters discuss the use of spikes over the automotive body to reduce the drag force and parametric study of the grinding process for the energy conservation, respectively.

Further, the application of nanomaterials is discussed in the eighteenth and twentieth chapters in different applications; the former used the nanomaterials to improve the insulation characteristics of electrical components under the rainy conditions, while the latter presented an extensive review on the employment of the nanomaterials to prepare nanohybrid-embedded polymer-based nonvolatile resistive switching board. To find the energy losses and maximum available energy in a thermal power plant, an exergy analysis of 660 MW thermal power plant is presented in the twenty-first chapter. Another study on the solar energy is presented in the twenty-second chapter, where a numerical analysis is conducted to investigate the effect of different jet geometries of air heater to attain the maximum heat transfer. The twenty-third chapter presents a very important work on the burning of agriculture residue in the open area and quantified the data to emphasize the hazardous nature of the pollutants emitted from the burning in the atmosphere of nearby area. The twenty-fourth chapter discusses the effectiveness of thermal spray coating to prevent the erosion and corrosion due to working conditions at very high temperature. The twenty-fifth chapter presents an insight about the analysis of HVAC system for buildings.

A novel process is presented in the twenty-sixth chapter to generate the hydrogen and syngas using the thermochemical splitting process of water using a unique solar thermochemical cavity reactor (STCR). In the continuation, the twenty-seventh chapter presents the optical and thermal analysis of STCR under the flow and no-flow conditions and the distribution of solar flux in the reactor is investigated. The use of bio-diesel reduces the depletion of conventional fossil fuels, but the effect of such mixture of bio-diesel and conventional fuel on various components of IC engine is a thrust area to make the bio-diesels commercially adoptable. The last chapter of this book presents a review on similar studies where the effect of bio-diesels is studied on the lubrication system of IC engine and raised a need to modify the properties of lubricating oil to make the use of bio-diesels sustainable in the future.

This book covers novel analysis, experimental findings, mathematical models and numerical simulations on recent developments on energy systems and future challenges of energy generation, storage, demands and supply as well as development in sustainable technologies. The editors have enough confidence that this book will be very helpful for the readers and provide a benchmark for future developments and research scopes in the energy sectors.

Indore, India — Ritunesh Kumar
Subang Jaya, Malaysia — Adarsh Kumar Pandey
Jaipur, India — Ravi Kumar Sharma
Phagwara, India — Gavendra Norkey

# Contents

# Contents

# About the Editors

**Dr. Ritunesh Kumar** is working as an Assistant Professor in the Department of Mechanical Engineering, Indian Institute of Technology Indore since October 2009. He received Ph.D. (Thermal Engineering), from Indian Institute of Technology Delhi (IITD), New Delhi, India, (2008) for his work on desiccant cooling. His focus is on refrigeration & air-conditioning, renewable energy, and heat transfer. He did his M.Tech. (Thermal Engineering) from Indian Institute of Technology Roorkee, Roorkee, India, (2002) and B.Tech. (Mechanical Engineering) from M. M. M. Engineering College, Gorakhpur, India (2000). Before joining IIT Indore, he was working for Tata Consulting Engineers Limited Mumbai. He has over 25 high impact articles and he is also a reviewer of many reputed journals.

**Dr. Adarsh Kumar Pandey** is associated with Research Centre for Nano-Materials and Energy Technology, Sunway University Malaysia. He obtained his Post Graduate Diploma in Rural (Renewable) Energy Technology and Doctorate degree in the field of Energy from SMVD University, India, and also earned two post-doctoral fellowships. He has also guided many Ph.D. and Master's theses. His major areas of research interests include phase change materials, nano-enhanced phase change materials (NEPCMS), hybrid solar photovoltaic (PV/T) systems, dye sensitized solar cells, self cleaning of PV modules, energy and energy analysis of renewable energy systems. He has published nearly 100 papers in respected international journals. Dr. Pandey is a reviewer in many reputed journals and also shares responsibilities of editorial board member in various journals.

**Dr. Ravi Kumar Sharma** is currently working as an Associate Professor in the Department of Mechanical Engineering, Manipal University, Jaipur. He obtained his doctorate degree in the field of Heat Transfer Enhancement from University of Malaya, Malaysia. His major areas of research interests include thermal energy storage, phase change materials, nano enhanced phase change materials, shape stabilized phase change materials. He has published 12 papers in respected international journals. Dr. Ravi Kumar Sharma is a reviewer in many reputed journals like *Energy Conversion and Management*, *Applied Thermal Engineering*, *Journal of Thermal Analysis and Calorimetry* and many more.

**Dr. Gavendra Norkey** is currently working as an Associate Professor in the Department of Mechanical Engineering, Lovely Professional University, Phagwara, Punjab, India. He obtained his B.E. (Mechanical) and M.Tech. (Production Engineering) from the Madhav Institute of Technology and Science, Gwalior (M.P.), and Ph.D. (Mechanical Engineering) from the Jaypee University of Engineering and Technology, Guna (M.P.). His major areas of research interests include production (manufacturing), advanced machining processes, laser materials processing, design of experiments, micro-manufacturing processes. He has published 24 research papers in different national and international journals and conferences of repute. He has guided five M.Tech. and one Ph.D. thesis and two candidates are pursuing Ph.D. with him presently. He is the life time member of Indian Laser Association (ILA) and International Association of Engineers (IAENG). He is the reviewer of reputed journals. He has 18 years of teaching and 1.5 years of industrial experience to his credit.

# Experimental Study on Rice Straw-Based Thermal Insulation

**Ranbir Singh Rooprai, Vikrampreet Singh, and Talvinder Singh**

**Abstract** Technological developments have a great effect on the environment. As a result, it is of much importance to analyse the impact of environment on our lifestyle. As aspects of life, people choose to live in environment-friendly houses. We can utilize our own resources to build these kinds of innovative structures. One such kind of renewable resource has been presented here in this work. India is an agriculture-based country. Straws are produced after harvesting the rice paddy. These strokes are either fired or plucked back to soil to create toxic gases as a black cloud of carbon that causes different diseases for live entities. In addition, burning these straws does not necessarily consume all materials as rapidly as possible, filling the air with gaseous and particulate contaminants and the soil. Ploughing them into the field is also not an efficient way of disposing them. Straw housing could be one of India's best approaches, where agriculture remains the main source of revenue, and straw production is enormous. Straw bale construction can be one of the best choices because of the large cost-effective attributes of renewable energy resources with good thermal strength, low weight and eco-friendly design. We could use multiple manufacturing processes, such as load bearing stroke or non-charge bearing construction, depending upon the needs. Straw bale design may be one of the best options for constructing a stable economic building.

**Keywords** Straw · Thermal insulation · Eco-friendly

---

R. S. Rooprai (✉) · V. Singh · T. Singh
Chitkara University Institute of Engineering and Technology, Chitkara University, Rajpura, Patiala, Punjab, India
e-mail: ranbir.rooprai@chitkara.edu.in

V. Singh
e-mail: vikrampreet.singh@chitkara.edu.in

T. Singh
e-mail: talvinder.singh@chitkara.edu.in

 
R. Kumar et al. (eds.), *Recent Trends in Thermal Engineering*, Lecture Notes in Mechanical Engineering, https://doi.org/10.1007/978-981-16-3132-0_1

# 1 Introduction

In today's time, energy saving is the utmost priority for each sector. Most energy is consumed due to heating and cooling loads in the construction sector. Insulation materials can be used in the constructions to save the energy and provide comforts as well. Different kind of materials needed to fulfil the requirements at different environment conditions [1–3]. Nanotechnology and nanocomposites have also been well known to be an important factor in improving certain features in this area. Selection of material has been made based on the properties such as thermal conductivity, sound insulation, waterproofing, fire-resistance and the effect of environment on the material [4–6]. Additional considerations such as availability and cost are also included in this approach for the selection of material. This emphasized the seeking of new materials and encouraged research institutions around the world for finding new materials that could be available on the market at a reasonable price [7]. At present, some organizations classified the traditional and alternative materials according to consumer requirements and history. Presently, wool and synthetic foams are the most common insulation materials used with less natural contributor [8]. The major problem of the majority of conventional materials is the high energy consumption during their manufacturing process. Moreover, these products are being manufactured from non-renewable petroleum-extracted products. This is the reason why alternative materials particularly natural materials are encouraged [9]. Recently, environmental concerns such as global warming and pollution have been a vital issue. With respect to global warming, this intake raises the amount of carbon dioxide in the atmosphere, which absorbs heat and forms global warming, because of power consumption in the manufacturing and the construction industries. Agriculture wastes pollute the atmosphere and have harmful effects on the creatures. The unsaturated epoxy resin and rice straw have been analysed and tested on the basis of energy savings and environmental isolation composite. For several factors such as low production expenses, safety in the atmosphere and low thermal conductivity, rice straw is selected [10, 11].

# 2 Ingredients

The ingredients used in this work are as follows.

## 2.1 *Rice Straw*

Rice straw is the rice plant's vegetative component, cut during or after grain harvest. Before the next ploughing, it can be burned and left on the farm, plucked as a soil enhancement or as livestock feed [12]. It is natural fibre. It has been used in this

**Fig. 1** Rice straw

work due to easy availability, low cost, low thermal conductivity and sustainable resources worldwide, and Fig. 1 represents the rice straw.

## 2.2 *Epoxy Binder*

Epoxy resin, also known as polyp epoxides, is a class of epoxy prepolymers and re-active polymers. It is used as a binder. Epoxy binder is used in Arcane Industries as epoxy paint application for all kind of surfaces [13]. Epoxy binders and resins are a group of highly adaptable oligomers and polymer that undergo thermosetting or treatment in all surfaces (collectively referred to as hard-wearing epoxy resins).

## 2.3 *Hardener*

Hardener is a multi-type mixture component. Hardener is used to vary the setting time of the composite. Some blends actually use a hardener to improve the strength of the blend when it is calculated. A hardener is used as a cure part in other mixtures. In a chemical reaction during the mixing phase, a hardener may be either a reactant or a catalyst. It is often known as accelerator as well. Epoxy resins are cured using hardeners. However, the simple addition of a hardener to an epoxy resin cannot lead to a rapid cure of the epoxy mixture.

## 2.4 *Cement*

Cement is a substance used to create, set and harden. It is an adhesive used to bind sand and aggregates into the most flexible materials of building. Cement could never be used alone but instead to bind together sand and gravel [14]. Cement mixed with a fine aggregate produces mortar for maceration or cement. It is referred to very fine-powdered material composed primarily of chalk (calcium), cement or

**Fig. 2** Cement

clay (silicone), bauxites (aluminium) and iron ore. It is common for shells to be used including marl, shale, coal, blast furnace slag and slate (Fig. 2).

## 2.5 *Fly Ash*

Fly ash is a combustion medium for fuel substance consisting of the particles powered from coal-fired boilers, and the flue gases are fly ash or flue ash also known as pulverized fuel ash. Fly ash is a strong mineral filter in hot mix asphalt (HMA) and increases flow ability in fills that are able to flow and grouts with a distinctive spherical form as well as particles distribution [15]. Fly ash is used in plain cement concrete (PCC) applications to increase strength.

## 2.6 *Coarse Sand*

Sand is a granular aggregate composed of finely separated mineral and rock particles. The coarser sand is our cement-filled sand that is washed and shielded to a greater grain than our fine-washed sand (masonry sand). It is much finer than gravel and has better strength. For the manufacture of ready-mixed concrete, coarsely sand is used for aggregate, water and cement (Fig. 3).

**Fig. 3** Coarse sand

### 2.7 *Sodium Hydroxide (NaOH)*

Sodium hydroxide is an inorganic compound chemically represented by the NaOH, also named as lye and caustic soda. This is a white solid ionic substance compound of Na+ and OH− ions. It is widely used for cleaning solutions but is also used in manufacturing items such as textiles and papers. Figure 6 represents NaOH. This product can be very dangerous and can damage the materials very badly. People should avoid direct contact with sodium hydroxide as it can lead to extreme skin burns and serious tissue damage when they are inhaled. NaOH is generally used in processes like oil refining, chemical manufacturing, metal processing and water treatment as well (Fig. 4).

## 3 Equipments Used

### 3.1 *Weigh Machine*

The machine weighs the weight of ingredients needed for making samples. A weighing scale is a tool to calculate weight or mass. It is referred to as mass scales, weight scales, mass balances, equilibrium or balance scale as well. Figure 5 shows the weigh machine. It is commonly used in the market since any items are mass-produced and packed.

**Fig. 4** Sodium hydroxide

**Fig. 5** Weigh machine

**Fig. 6** Mould

## 3.2 Moulds

A mould is a hollowed block of metal. The fluid hardens or sets in the form of the mould. A casting is produced with the help of a mould. Sheet of steel (5 mm thickness) is required to make the mould of required size. Mould size is of $50 \times 50 \times 50$ mm as shown in Fig. 6.

## 3.3 Compression Moulding Machine

The mixture of composite is filled in the moulds and then compressed by the machine, so that it can take shape of mould. A high strength or stepping component closes the mould and forces on the material to make contact with all areas of the mould and retains the heat and pressurization until the moulding material has cured.

## 3.4 Compression Testing Machine

Samples should have sufficient compressive strength. After the samples are ready, we can check the compressive strength of these samples. The compression strengths of the samples are tested on the universal testing machine.

## 3.5 Breaded and Un-breaded Mixture Samples

The ingredients discussed above have been taken in the ratio as shown below: Epoxy 15% + hardener 7% + fly ash 14% + cement 10% + sand 30% + rice straw 24%.

## 4 Experiment Work

The rice straws were dipped in the solution of NaOH and water (5 mol/l), for treatment of rice straw. Dry straws were bounded into a pattern, so that they can be easily put in the composite. The straws (shown in Fig. 7) were left for one day to dry out. The ingredients were taken in the pan and mixed. Breaded and un-breaded samples were made. In breaded samples, the rice straws were put in composites, layer by layer while in un-breaded samples, the rice straws (cut into small pieces) randomly put in the composite. The samples produced have been shown in Fig. 8. The composites then compressed in moulding machine, so that extra moisture can move out of the mixture, and it can dry easily and can take the shape of the mould. After one day, the samples were taken out of the moulds and kept out in sunlight. Compression test is done on all samples of breaded and un-breaded types.

**Fig. 7** Patterns of rice straw

**Fig. 8** Samples

## 5 Observations

The compression test has been performed on universal testing machine (UTM) shown in Fig. 9. The surface area of all samples kept 2500 mm$^2$. When the breaded samples kept under the load, it was able to bear up to 21 kN of force. After this load, it breaks slowly. As in the breaded samples, the rice straws were put in an order of layers, and hence, sudden breaking did not occur in the sample. When un-breaded samples were put under the load, it could only bear up to 18 kN of load. As in this sample, the rice straws, first, cut into small pieces and then put in the sample randomly. Thermal conductivity is the property of a material for conducting heat. It is evaluated primarily in terms of Fourier's law for heat conduction. Heat transfer occurs in thermal conductivity materials with lower concentrations than in high thermal conductivity materials. Thermal conductivity of rice straw is low.

### *5.1 Compressive Strength*

For 20% breaded, maximum load = 21.25 kN

Surface Area = 50 × 50 mm$^2$ = 2500 mm$^2$

$$\text{Strength} = \frac{21.25 \times 1000}{2500} = 8.5\,\text{N/mm}^2 \tag{1}$$

For 25% breaded, maximum load = 20 kN

Surface Area = 50 × 50 mm$^2$ = 2500 mm$^2$

**Fig. 9** Universal testing machine

$$\text{Strength} = \frac{20 \times 1000}{2500} = 8\,\text{N/mm}^2 \quad (2)$$

For 25% un-breaded, maximum load = 18 kN
Surface Area = 50 × 50 $mm^2$ = 2500 $mm^2$

$$\text{Strength} = \frac{18 \times 1000}{2500} = 7.2\,\text{N/mm}^2 \quad (3)$$

## 5.2 *Thermal Conductivity*

Thermal conductivity is calculated based on the Fourier's law of heat conduction. The direction of heat flow is perpendicular to the straw fibres.

$$\text{Fouriers law, } Q = -k.A.\frac{\mathrm{d}T}{\mathrm{d}S}$$

where

$Q$ heat transfer
$K$ thermal conductivity
$A$ 2500 $mm^2$
$\mathrm{d}T$ temperature difference
$\mathrm{d}x$ 50 mm

$Q$ heat transfer
$K$ thermal conductivity
$A$ 2500 $mm^2$
$\mathrm{d}T$ temperature difference
$\mathrm{d}x$ 50 mm

Thermal conductivity found to be 0.051 W/mK when the heat flow is perpendicular to straw bale.

# 6 Conclusion

From the observations, it is concluded that the load bearing capacity for the breaded mixture samples is higher than the un-breaded mixture. Further, it has been seen that as the percentage is increased in the breaded mixture then the load bearing capacity also decreases. Therefore, depending upon the requirements, we can go for the amount of percentage in the breaded samples. The thermal conductivity is found to be low as compared to the conventional materials. Thus, it can be used as an insulation material in an eco-friendly way.

## References

1. Panyakaew S, Fotios S (2008) Agricultural waste materials as thermal insulation for dwellings in thailand: preliminary results. In: 25th conference on passive and low energy architecture, Dublin
2. Mounika M, Ramaniah K (2012) Thermal conductivity characterization of bamboo fiber reinforced polyester composite. J Mater Environ Sci
3. Binici H, Aksogan O (2015) Mechanical, thermal and acoustical characterizations of an insulation composite made of bio-based materials. Sust Cities Soc
4. Binici H, Eken M, Kara M (2013) An environment-friendly thermal insulation material from sunflower stalk, textile waste and stubble fibers. In: International conference on renewable energy research and applications
5. Gustavsson L, Sathre R (2005) Variability in energy and carbon dioxide balances of wood and concrete building materials. Build Environ
6. Paiva A, Pereira S (2011) A contribution to the thermal insulation performance characterization of corn cob particle boards. Energy Build
7. Morela JC, Mesbah A (2000) Building houses with local materials: means to drastically reduce the environment impact of construction. Build Environ
8. Cantorski da Rosa L, Santor CG (2015) Use of rice husk and sunflower stalk as a substitute for glass wool in thermal insulation of solar collector. J Cleaner Prod
9. Korjenica A, Petranek A (2011) Development and performance evaluation of natural thermal-insulation materials composed of renewable resources. Energy Build
10. Panyakaew S, Fotios S (2011) New thermal insulation boards made from coconut husk and bagasse. Energy Build
11. Pinto J, Paiva A (2011) Corn's cob as a potential ecological thermal insulation material. Energy Build
12. Mussoline W, Esposito G (2013) The anaerobic digestion of rice straw: a review. Environ Sci Technol
13. Glaskova TI, Guedes RM (2007) A comparative analysis of moisture transport models as applied an epoxy binder. Mech Comp Mater
14. Tazawa E, Miyazawa S (1995) Influence of cement and admixture on autogenous shrinkage of cement paste. Cem Conc Res
15. Ahmaruzzaman M (2009) A review on the utilization of fly ash. Progress in Energy and Combustion Science

# Experimental Exploration of Effect of Hydrogen Enrichment on the Performance and Emissions of Dual Fuel Diesel Engine Equipped with CRDI by Varying Injection Duration

**M. R. Dahake and D. N. Malkhede**

**Abstract** The hydrogen due to its carbonless structure is considered as a potential supplement fuel in near future for dual-fuel Internal Combustion engines. It reduces the burden of energy imports and reduces carbon containing tailpipe emission, thereby protecting the environment. Hydrogen has inimitable characteristics because of carbonless structure which is considered as better alternative fuel compared to other available options, for example, liquefied petroleum gas, compressed natural gas, etc. In the presented study, investigation was conducted using hydrogen gas in dual-fuel method in single-cylinder CI engine. Hydrogen fuel was inoculated in intake manifold for different injection duration while injecting diesel into combustion chamber directly. The performance of engine is compared with baseline diesel performance at varying injection duration. Experimental observations demonstrated the performance improvement and exhaust emissions using hydrogen enrichment technique. The brake thermal efficiency observed to be improved by 3.17%, and brake specific energy consumption reduces by 10.81% at fully loaded condition for hydrogen gas injection duration of 6 ms as that of baseline diesel performance. Improvement is seen in performance parameters as well as in emissions also, hydrocarbon reduces by 68.18%, and carbon dioxide reduces by 43.33% at full load condition with same injection duration of 6 ms. It was experiential that due to homogeneous mixing of hydrogen with air leads to complete combustion of fuels with lesser emissions. The current study proved that hydrogen enrichment is a potential technology which could be used in compression ignition engines to improve performance and lessening emissions without any major modification in hardware of basic diesel engine.

**Keywords** Hydrogen enrichment · Injection duration · Dual-fuel engine · Performance and emissions

M. R. Dahake (✉) · D. N. Malkhede
College of Engineering, Pune, India



R. Kumar et al. (eds.), *Recent Trends in Thermal Engineering*, Lecture Notes in Mechanical Engineering, https://doi.org/10.1007/978-981-16-3132-0_2

## 1 Introduction

The conventional petroleum-based fuels are being widely used for more than a century as the main source of fuels for transportation and power generation. But the automobile sector is fronting the threat because of interdependence and surety of petroleum fuels availability and environmental problems. Alternate fuels can address the said problems up to some extent, but almost all in these categories are hydrocarbon-based fuels which emit $CO_2$, known as greenhouse gas (GHG). Apart from $CO_2$ emission, other emissions such as HC, CO and $NOx$ which are discharged from the automobiles lead to air pollution [1, 2]. Environmental concerns have motivated legislative actions by government around the world for reducing emissions and improving economy of fuel. In view of reducing the harmful emissions, various alternate fuels are investigated for the substitute of conventional petroleum-based fuels. The probable alternate fuels which can replace them are mostly compressed-natural gas (CNG), liquefied natural gas (LNG), liquefied petroleum gas (LPG), hydrogen ($H_2$), alcohols, vegetable oils, producer gas, etc. Hydrogen can be well-thought-out as a potential alternate fuel due to its carbon less structure, renewable, less polluting and clean burning, nontoxic, odourless and leads to fuels complete combustion [3, 4]. Carbonless structure of hydrogen leads to reduction of CO, $CO_2$ and HC emissions. Diesel engines could be used with hydrogen with little or without any major modification in engine in dual-fuel mode. There are many exceptional aspects of hydrogen that enormously influences technological developments of hydrogen IC engines which are compared with other fuels and are recapitulated as given in Table 1.

Due to the characteristics of hydrogen gas such as wide flammability range, small quenching distance, flame velocity, low density and high auto-ignition temperature, hydrogen is being focused as an alternate fuel for diesel engine in particular. The carbonless structure being the most substantial among all peculiar feature of hydrogen compared to other hydrocarbon fuels. The rapid combustion can be achieved because of higher burning velocity. The flammability limit of hydrogen allows diesel engine to run with a very large series of air–fuel ratio [3]. Minimum ignition energy permits to run on lean mixtures to hydrogen engine

**Table 1** Combustion characteristics of hydrogen [4–7]

| Particular | Diesel | Gasoline | Hydrogen |
|---|---|---|---|
| Self-ignition temperature (°C) | 257 | 260–460 | 585 |
| Limits of flammability (vol % in air) | 0.7–5 | 1.4–7.6 | 4–75 |
| Stoichiometric ratio (A: F) | 14.5 | 14.6 | 34.3 |
| Density (kg/m$^3$) | 833–881 | 721–785 | 0.0838 |
| Calorific value (kJ/kg) | 42,500 | 43,900 | 119,930 |
| Flame velocity (cm/s) | 30 | 37–43 | 265–325 |
| Minimum ignition energy (mJ) | – | 0.24 | 0.02 |
| Diffusion rate in air (cm$^2$/s) | – | 0.08 | 0.63 |

ensuring prompt ignition. The lighter hydrogen density has an advantage of diffusing into the atmosphere easily compared to air and has high energy–weight ratio among all available fuels. High cylinder pressure rise rate can be achieved because of higher flame velocity. A larger compression ratio in hydrogen internal combustion engine is possible by reason of high auto-ignition temperature of hydrogen. But, some means of ignition are required inside the combustion chamber as compression alone is not sufficient to achieve prompt ignition of hydrogen [8].

Numerous investigates have been conducted on hydrogen engines in different conditions with different techniques. Some of the researchers used hydrogen in diesel and petrol engines in a dual-fuel method of operation, some of the researchers used hydrogen with EGR of hot or cooled even tried with superchargers and turbochargers, and few used hydrogen-rich gases and hydroxy gas which is produced by water electrolysis process. Hydrogen was tried with biodiesel also in CI engines. Tsujimura et al. [9] have studied combustion technologies development of hydrogen in IC engine with high efficiency with lower emission focussing mainly on dual-fuel combustion technology which includes diesel fuel and hydrogen for study. Yadav et al. [10] have presented the combustion investigation of CI engine attaching exhaust gas recirculation system enriched with hydrogen and concluded the enhancement in performance and emissions parameters by means of hydrogen in dual-fuel method as an alternate fuel. Christodoulou et. al. [11] has analysed the combustion effect in diesel engine by simultaneously adding nitrogen and hydrogen and concluded that $H_2 + N_2$ addition has a disadvantage of fuel consumption. Manu et al. [12] have experimentally depicted the working of diesel engine with enriching hydrogen with incoming air in the intake manifold using an on-board dry cell electrolyser. Imran et al. [13] have studied engine performance as well as its emissions contours by RME as a pilot fuel and diesel for the complete operation of hydrogen-fuelled CI Engine. Saravanan et al. [14] have researched on effect of hydrogen injection in intake port at the same time as injecting diesel into the combustion chamber directly. They concluded from their experimental analysis that dual-fuel technique with hydrogen in diesel engine enhances the performance along with its exhaust emission except NO*x*. Arya et al. [15] have reviewed tyre pyrolysis oil which is extracted by pyrolysis technique from waste tyre as an alternative fuel for diesel engine and stated that it can be used to petroleum fuel consumption due to their increasing prices with many benefits such as carbon dioxide recycling, reduced environmental pollution which otherwise by burning waste tyres and value-added chemicals. Sharma et al. [16] have focussed on regeneration of used lubrication oil which otherwise will be thrown in to environment, and then, blending of distilled lubricating oil was blended with jatropha biodiesel to explore the performance along with emission characteristics for different blends. The study suggested the best blends as 80% jatropha and 20% distilled utilized lubrication oil from the behaviour of engine run.

## 2 Methodology

During experimentation work, hydrogen gas was inducted into intake manifold which is stored at high pressure (150 bar) which can be decreased to an injection pressure of 2.5 bar by two stage pressure regulator from 150 to 10 bar and then from 10 bar to injection pressure. The hydrogen was permitted to flow via the flame arrestor to avoid bursting in hydrogen supply and then enters into the flame trap which is used to extinguish flame inside the supply line in case of backfiring. Hydrogen flow rate was measured with Emerson make digital gas flow meter. Then, the hydrogen was permitted to flow inside the inlet manifold by solenoid-type hydrogen gas injector with the help of Open ECU system. Performance of the engine is measured through the ICEngineSoft software, and AVL DiGas-444 emission analyzer was used to measure emissions parameters. Smoke opacity was being counted with AVL-smokemeter. Experimental results obtained at different hydrogen injection duration at different loading conditions were compared with baseline performance.

## 3 Test Engine

Figure 1 shows the experimental set-up with hydrogen supply and safety system used for experimental work. The experimental set-up comprises four-stroke single-cylinder CRDI diesel engine attached with eddy current dynamometer for loading purpose. All necessary instrumentation and devices are provided for combustion analysis and in-cylinder pressure measurements. The experimental set-up allows complete investigation of engine performance, emission and combustion characteristics at varying hydrogen injection duration. Table 2 shows the specifications of described engine.

## 4 Error Analysis and Determination of Uncertainity

All experimental investigations possess certain uncertainties whatever the type of instruments used may be by reason of fixed/random errors. Uncertainity analysis is needed to prove accurate measurement of experimental data. Due to repeatability of fixed errors, it may be effortlessly accounted for measurement of true value where random errors can be estimated analytically. The average uncertainty of some measured and calculated parameters is shown in Table 3.

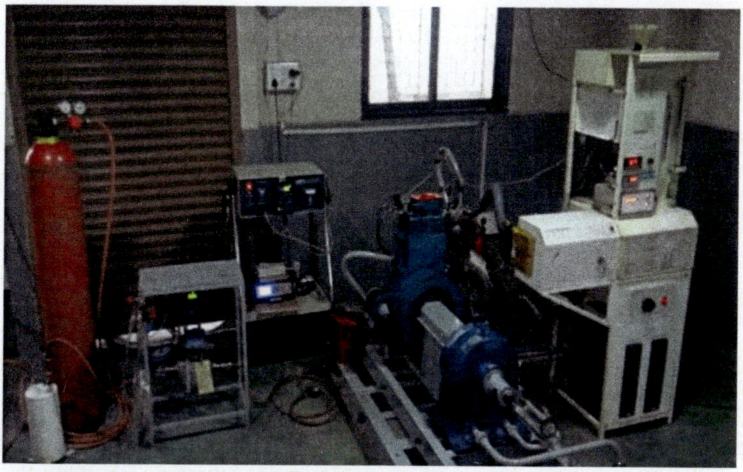

**Fig. 1** Experimental set-up used with hydrogen supply and safety system

**Table 2** Diesel engine specification used for experimentation

| Engine parameter | Value |
|---|---|
| Bore/stroke (mm) | 87.5/110 |
| Number of cylinder | 01 |
| Engine displacement (cc) | 661 |
| Engine power rating (kW@ 1500 rpm) | 3.5 |
| Connecting rod length (mm) | 254 |
| Compression ratio | 12–18 |

**Table 3** Average uncertainity of calculated and measured parameters

| Parameter | Uncertainity % |
|---|---|
| Brake power | 4.7 |
| Speed | 5.2 |
| Flow rate of diesel | 3.2 |
| Flow rate of air | 3.8 |
| Flow rate of hydrogen | 2.4 |
| Hydrocarbons | 3.8 |
| Oxides of nitrogen | 4.2 |
| Smoke | 4.2 |

## 5 Result and Discussions

In current experimentation work, hydrogen was supplied along-with in coming air through intake manifold with different injection duration from 2 to 6.4 ms at step of 2 ms. The performance parameters and emissions at every step of enrichment were compared with baseline diesel performance.

### 5.1 *Brake Thermal Efficiency*

The change in brake thermal efficiency with regard to engine loading was depicted in Fig. 2 in which improvement in it is observed at all loading conditions with all hydrogen enrichment percentage. This is because of improvement in combustion process due to hydrogen fuel which improves mixing process, shortens the combustion duration and its better combustion charecteristics. The hydrogen enrichment at 6 ms gives highest efficiency (28.12%) as that of neat diesel (24.95%) at fully loading condition. The knocking tendency limits the ratio of hydrogen enrichment at 6.40 ms at 75% load as shown in figure.

### 5.2 *Brake Specific Energy Consumption*

The effect of brake specific energy consumption with regard to engine load is seen in Fig. 3. BSEC decreases with the rising in load as hydrogen enrichment percentage rises at constant speed at all loads. This is because improved integration of

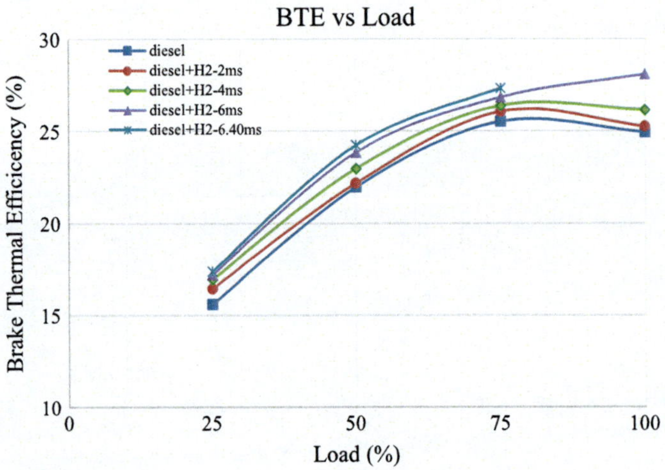


**Fig. 2** Variation of brake thermal efficiency at different injection duration

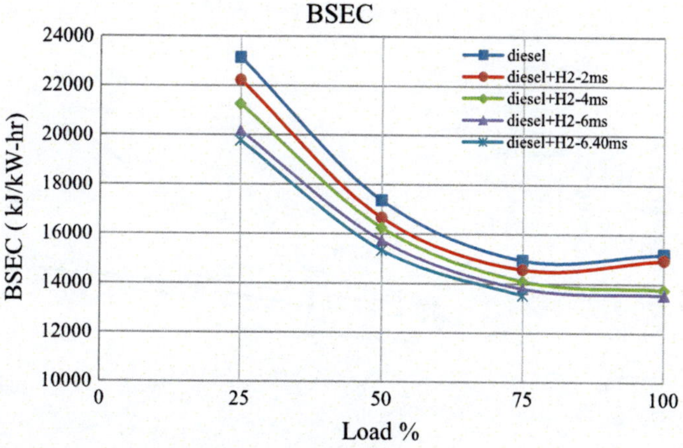


**Fig. 3** Variation of BSEC at different injection duration

fuel along with air results in improving combustion process, and high CV decreases fuel consumption to produce unit power. The lowest BSEC of 13558 kJ/kW-hr is obtained for 6 ms at fully loaded condition of engine as that of diesel of 15202 kJ/kW-hr.

## 5.3 *Hydrocarbons*

The variation in hydrocarbons with load is shown in Fig. 4. It is found that a hydrocarbon reduces with increasing hydrogen enrichment at all loading conditions except for 2 ms case at 25% load only. Higher burning velocity of hydrogen reduces HC and enhances the combustion process. The HC emissions reduce by larger extent as reason of carbonless structure of hydrogen fuel. The lowest value 7 ppm for hydrogen addition was recorded at full load against 21 ppm for 6 ms compared to neat diesel operation.

## 5.4 *Carbon Monoxide*

Figure 5 portrays the deviation of carbon monoxide with regard to engine load at different hydrogen enrichment percentage. It is seen that at full loading condition, CO level of hydrogen-operated engine is lower compared to neat diesel operation due to carbonless structure of hydrogen fuel. The lowest CO was observed as 0.28% by vol at higher enrichment ratio compared to neat diesel of 0.41% by vol. The absence of carbon is the main reason for lessening in CO emission.

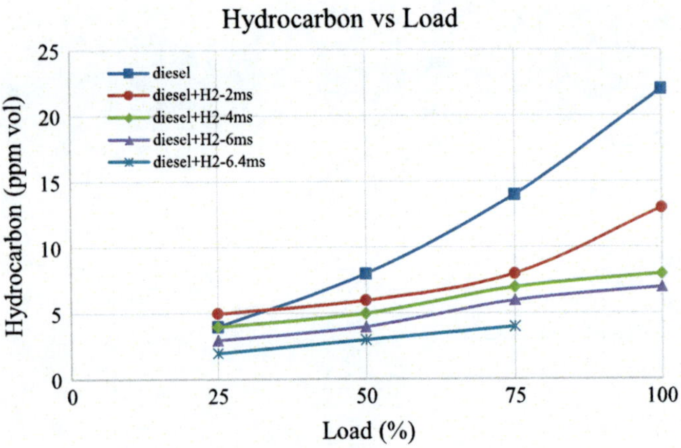


**Fig. 4** Variation of hydrocarbon at different injection duration

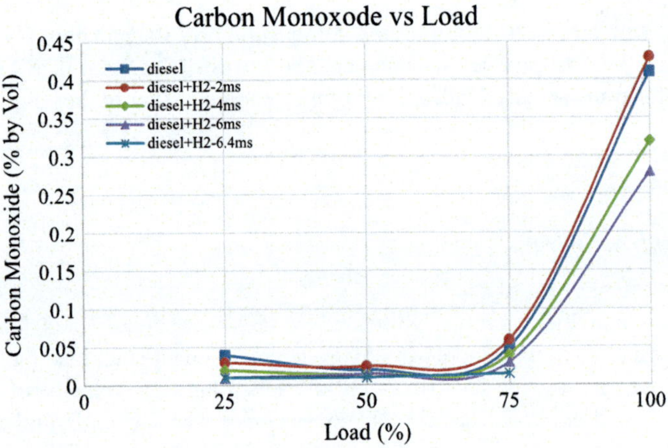


**Fig. 5** Variation of CO at different injection duration

## 5.5 *Carbon Dioxide*

The deviation of carbon dioxide with respect to engine load is shown in Fig. 6. The $CO_2$ percentage was found to be less with hydrogen-enriched engine in comparison with neat diesel operation. Carbonless structure of hydrogen fuel is the significant cause for lessening in $CO_2$ concentration. The lowermost level of $CO_2$ was observed 1.7% by volume which is achieved with 6 ms as that of neat diesel 3% by volume at fully loaded condition. It has been also observed that the $CO_2$ increases at lowest hydrogen enrichment ratio compared to that of neat diesel.

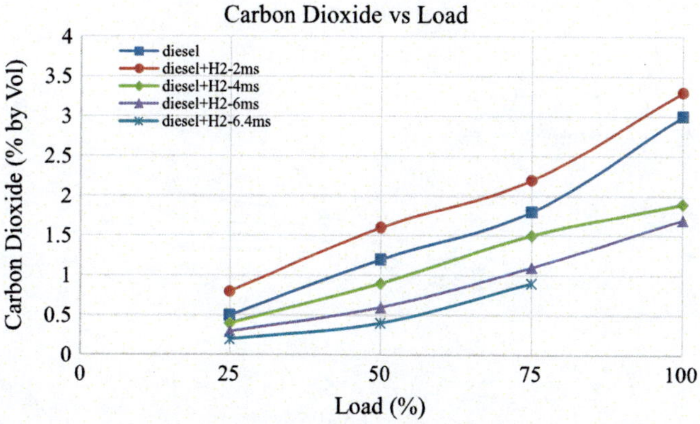


**Fig. 6** Variation of $CO_2$ at different injection duration

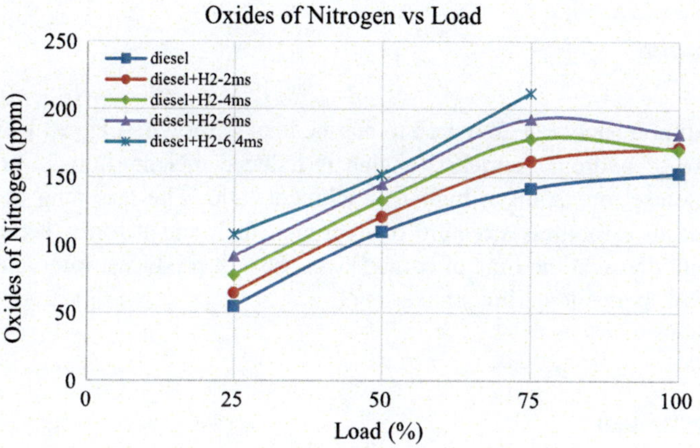


**Fig. 7** Variation of oxides of nitrogen at different injection duration

## 5.6 *Oxides of Nitrogen*

Figure 7 depicts the increase in oxides of nitrogen with regard to engine load at different hydrogen blending percentage. It is found that NO*x* increases up to 75% of load and then reduces in comparison with neat diesel-operated condition for higher two enrichment ratios considered. Major cause in the increase of NOx was higher combustion temperaure due to high calorific value of hydrogen and high oxygen concentration. The minimization is mainly because hydrogen-operated engines run at leaner equivalence ratios. As enrichment of hydrogen rises, the lean limit of combustion increases which leads to reduction in peak combustion temperature.

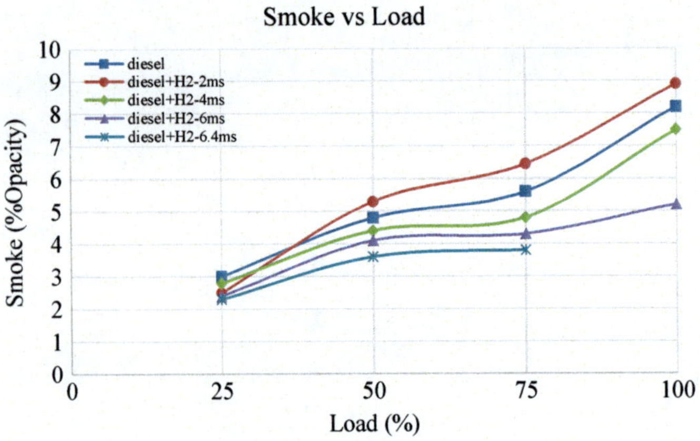


**Fig. 8** Variation of smoke at different injection duration

### 5.7 *Smoke*

The deviation in smoke with regard to engine load is shown in Fig. 8 which shows that smoke of hydrogen-enriched engine is reduced meaningfully in comparison with neat diesel operation at higher enrichment ratio. The lessening in smoke is mainly due to carbonless structure of hydrogen fuel, and it forms better uniform mixing with diesel at the time of combustion. Also, it produces water and does not produce PM, hence lowering smoke level.

## 6 Conclusion

The dual-fuelled single-cylinder CI engine equipped with CRDI was operated successfully up to 6 ms of injection duration, and based on experimental results, the following conclusions are made:

- The brake thermal efficiency found to be increased by 3.17% at full load because of enhancement in combustion process due to hydrogen fuel as it improves mixing process, shortens the combustion duration and provides better combustion characteristics.
- Brake specific energy consumption reduces by 10.81% at fully loaded condition for hydrogen gas injection duration of 6 ms. This is because of improved mixing of fuel with air leads to improving combustion process as well as high CV decreases fuel consumption to produce unit power.
- Hydrocarbons reduces by 68.18%. Higher burning velocity and carbonless structure of hydrogen reduce the HC and enhance the combustion process.

- CO and $CO_2$ both reduce by 31.70% and 43.33%, respectively, at full load condition at injection duration of 6 ms mainly because of the carbonless structure of hydrogen fuel.
- Oxides of nitrogen increase up to 75% of load and then reduce as compared to that of neat diesel-operated condition at higher two enrichment ratios considered. The minimization is mainly because hydrogen-operated engine runs at leaner equivalence ratios.

In this current experimental work, the hydrogen gas enrichment techniques result in better combustion and reduced emissions and have potential of getting maximum benefit in engine performance. But the increase in NO$x$ could not be prevented which could be further minimized by exhaust gas recirculation approach up to few amount. Hydrogen enrichment is a prospective technology which can be used in diesel engines for inflexible emission conformity without any major modification in basic hardware of diesel engine.

## References

1. Srinivas J, Saravanan N, Das L (2015) An experimental study of turbocharged hydrogen fuelled internal combustion engine. SAE Tech Pap 2015–26–0051. https://doi.org/10.4271/2015-26-0051
2. Saravanan N, Nagarajan G (2008) An experimental investigation of hydrogen–enriched air induction in a diesel engine system. Int J Hydrogen Energy 33(2008):1769–1775
3. Saravanan N, Nagarajan G, Dhanasekaran C, Kalaiselvan KM (2007) Experimental investigation of hydrogen port fuel injection in DI diesel Engine. Int J Hydrogen Energy 32(2007):4071–4080
4. Murat C, Huseyin K (2012) Hydrogen use in internal combustion engine—a review. Int J Automot Eng Technol 1(1):1–15
5. Saravanan N, Nagarajan G, Dhanasekaran C, Sanjay G, Kalaiselvan KM (2008) An experimental investigation on hydrogen as a duel fuel for diesel engine system with EGR technique. Int J Renew Energy 33:422–427
6. Yilmaz AC, Uludamar E, Kadir A (2010) Effect of hydroxy gas addition on performance and exhaust emissions in compression ignition engines. Int J Hydrogen Energy 2010:1–7. https://doi.org/10.1016/j.ijhydene.2010.07.040
7. White CM, Steeper RR, Lutz AE (2006) The hydrogen fuelled internal combustion engine: a technical review. Int J Hydrogen Energy 31(2006):1292–1305
8. Saravanan N, Nagarajan G, Dhanasekaran C, Sanjay G, Kalaiselvan KM (2008) Combustion analysis on a DI diesel engine with hydrogen in a dual fuel mode. Int J Fuel 87(2008):3591–3599
9. Tsujimura T, Suzuki Y (2017) The utilization of hydrogen in hydrogen/diesel duel fuel engine. Int J Hydrogen Energy 42(2017):14029–14039
10. Yadav VS, Soni SL (2015) Performance and combustion analysis of hydrogen fuelled C.I. engine with EGR. Int J Hydrogen Energy 40(2015):4382–4391
11. Christodoulou F, Megaritis A (2014) Experimental investigation of the effects of simultaneous hydrogen and nitrogen addition on the emission and combustion of diesel engine. Int J Hydrogen Energy 39(2014):2692–2702
12. Manu PV, Sunil A, Jayraj S (2015) Experimental investigation using an on-board dry cell electrolyzer in a CI engine working on dual fuel mode. In: Energy Procedia, 5th international

conference on advances in energy research (ICAER 2015), Mumbai, India, 90(2016), pp 209–216
13. Imran S, Emberson DR, Hussain A (2015) Performance and specific emission contours throughout the operating range of hydrogen-fueled CI engine with diesel and RME pilot fuels. Alexandria Eng J 54(2015):303–314
14. Saravanan N, Nagarajan G (2010) Performance and emission studies on port injection of hydrogen with varied flow rates with diesel as an ignition source. Int J Appl Energy 87 (2010):2218–2229
15. Arya S, Sharma A, Rawat M, Agrawal A (2020) TyrePyrolysis oil as an alternative fuel: a review. Mater Today Proc 28(4):2481–2484. https://doi.org/10.1016/j.matpr.2020.04.797
16. Sharma A, Gupta G, Agrawal A (2020) Utilization of waste lubricating oil as a diesel engine fuel. In: IOP conference series: materials science and engineering, vol 840, no 1, p 012015. https://doi.org/10.1088/1757-899X/840/1/012015

# Numerical and Analytical Investigation of Automotive Exhaust Gas Waste Heat Recovery Module Using Thermoelectric Generator

**Minesh Vohra, Mandil Sharma, and Sanjeev Kumar**

**Abstract** The study is based on numerical and analytical investigation of a thermoelectric generator system setup that can be mounted at the exhaust pipe of a 120 cc IC engine, i.e., to capture the amount of waste heat from the exhaust by generating temperature difference across the generator using thermoelectric conversion principle. This entire system consists of a heat absorber, thermoelectric generator module, and heat sink. The staggered fins arrangement has been introduced on both heat sink and heat absorber. The staggered pin–fin is placed into the duct to form the heat absorber, as duct installed at the exhaust of the engine so continuous heat passes through the heat absorber. The staggered fins on the heat absorber could increase the heat-exchange surface area, and by doing so, thermoelectric conversion increases. Simulation and analytical results are combined to verify power output results and the relation between the available power and expected power output.

**Keywords** Thermoelectric · Exhaust · Convective heat transfer coefficient · Waste heat recovery

## 1 Introduction

Olle et al. [1] system setup consists of a hot block sandwiched between two cold blocks. Thermoelectric generator of 80 × 80 mm with hot block is being heated using heater, whereas cold blocks make use of water flow to increase the rate of heat dissipation from the hot block. Different power input for heater provides variable results for temperature difference. Yu et al. [2] In his study has carries few assumptions, such as axial heat conduction being neglected and transverse heat dissipation dominating. The thermoelectric generator was used to conduct an experiment on the counter fluid flow of hot and cold regimes. Using Nusselt and

M. Vohra (✉) · M. Sharma · S. Kumar
Lovely Professional University, Phagwara, India
e-mail: minesh.15783@lpu.co.in



R. Kumar et al. (eds.), *Recent Trends in Thermal Engineering*, Lecture Notes in Mechanical Engineering, https://doi.org/10.1007/978-981-16-3132-0_3

Reynolds numbers at a certain place, Bi2Te3 is used to calculate the model's power output. Bai [3] Six different types of heat exchangers are tested numerically and experimentally. These designs with cavity, series, parallel, inclined, pipe-structured and holed plated fins are compared. Pressure drops are observed and found that series type of structure is effective as compared to others. Andrea et al. [4] Simulation of power generation considering the Peltier effect, Thomson effect and Joules heating of thermoelectric generator kept between hot and cold sides. Baskaya et al. [5] In their experimental and numerical study on condensing combi boiler, sixteen volt of potential difference is gained with 201 °C of temperature difference. Natural gas assists as a fuel to attain the efficiency of 4.5%, and it is suggested that there are possibilities to get much more temperature difference to produce maximum amount of power at max load condition. Lu et al. [6] Performance of thermoelectric generator for heat exchanger depends on design and material used for fins. Optimum thickness and spacing for fins increased the net power with suitable operating conditions. Huang et al. [7] Water as cooling medium for their concentric cylindrical thermoelectric system for automotive waste heat produced 6.62 A of current. Kim et al. [8] Study of automotive exhaust thermoelectric generator of gas at very high temperature for parallel fins through heat exchanger for variable parameters is simulated. Increase in the number of guide plates showed that the flow has decreased pressure drop. It was noted during the study that there is specific amount of increment in the heat flux at the surface with certain percentage of decrease in the pressure drop. Karri et al. [9] These case studies compare the fuel savings for stationary compressed natural gas engine and sports utility vehicle. Thermoelectric generator based on Bi2Te3 and Quantum well generated different amount of power. The increase in power between the Quantum well and Bi2Te3-based generators was about three times for the SUV and seven times for the CNG generator. Hsu et al. [10] Waste heat recovery from low temperature uses block to distribute the streams evenly. These streams later on induced the system to capture more amount of heat. Wu [11] In his research, the power outputs of various cycles such as Carnot, end reversible, and external reversible thermoelectric generators are compared. In an actual waste heat generation cycle, the thermoelectric generator underperformed. Moreover, the prediction of specific power for internal and external irreversibility effects was addressed in depth. Suzuki et al. [12] Numerical analyses for six systems of cylindrical multi-tube to increase the maximum power output were done. Compactness of design with reliable variation in maximum power output for specific boundary conditions helped in comparing with flat thermoelectric modules. Jadhav et al. [13] The system converts the squander heat from the four-stroke engine exhaust manifold into an electrical energy using thermoelectric generator. It consists of two thermoelectric generators connected thermally in parallel and electrically in series. Single cylinder engine with displacement of 92.20 cc is used which led to appreciable amount of increase in voltage. Kim et al. [14] Direct contact thermoelectric generator is experimentally tested for diesel engine. Variations in engine load, rotation speed and coolant temperature are kept in certain range for experiments. Efficiency of generator increased with large engine loads and high speed which lies in the range of approximately 1.0–2.0%. Conversion efficiency of generator showed increase of 0.25% with 10 K decrease in the coolant range. Yadav et al. [15] TEG on flexible fibber substrate is intended and demonstrated. They reasoned that an innovation can be effectively implemented in creating a flexible thermoelectric generator for waste heat recovery by vehicle and diesel/petroleum power plant. Min et al. [16] He created a novel tube-shaped thermoelectric module for the generation of electricity. Generator with his designed thermoelectric module showed an increase in power generation in comparison with conventional-type thermoelectric design. Jang et al. [17] Paper investigates the three-dimensional turbulent flow in a chimney used for venting flue gas from boiler or stove. Installing more thermoelectric generator resulted in a large amount of total power with quick saturation. Jeng et al. [18] Investigation of an actual four-stroke single cylinder of 35.8 cc I.C engine was used for the experimental setup for generation of electricity by exhaust of the engine. He concluded that power 2.5 W generated at a speed of the engine is 5400 rpm. Thermoelectric modules are used in utilizing the waste heat available at exhaust. There are many experimental studies conducted for different designs with varying number of fins and materials used as thermoelectric generator, but the numerical and analytical study has not been done on any of the module with the staggered fin orientation. Also, the compact system for the waste heat recovery was not yet formed which can be used in two wheelers for the generation of electricity based on the principle of thermoelectric generation. Hence, a compact design based on a staggered pin–fin arrangement with thermoelectric generator module has been proposed for simulation. The study deals with correlation of numerical and analytical results as per the proposed module which would be more effectively designed with low cost for the experimental progression. Hence, the major objective of the study is based on analytical and numerical investigation which gives us an accurate measure of the various parameters of TEG device which are listed below:

- Calculation on Carnot efficiency for the module with power loss during the waste heat recovery
- Variation of Carnot efficiency variation in thermal heat transfer coefficient based on increase in RPM and hence velocity at exhaust of an engine

## 2 Research Methodology

Basic purpose of following studies is to collect the data which is fulfilled by merging analytical and simulated information gathered from analysis done through computational fluid dynamics.

## 2.1 Numerical Investigation

Computational fluid dynamics (CFD) study begins with the modeling of desired geometry and meshing of modeled domain. Generally, geometry is made simple for CFD studies. Further number of fins modeled for the heat absorber sections are 82. Similarly, total number of fins at heat sink is 82. Meshing starts with division of the domain into small volumes where the equations are solved with the help of numerical methods. Analysis starts with the implementation of the initial and boundary conditions for whole domain. The mesh is checked, and quality is obtained. The analysis type is changed to pressure-based type. The velocity formulation is changed to absolute value and time interval to steady state condition. The following steps has been inculcate in the study, that are:

2.1.1 Modeling of device.
2.1.2 Domain finalization.
2.1.3 Meshing.
2.1.4 Analysis setup.
2.1.5 Initial and boundary conditions.

Boundary conditions are used according to the need of the experimental setup. The inlet and outlet conditions are defined as velocity inlet and pressure outlet. No slip condition is considered at walls. The analysis ran for tabulated inlet conditions. Assuming that, the experimentaiton work has been carried out under stationary position, so the convective heat transfer coefficient for air is determined at 1 m/s. This coefficient is used in simulation to find the heat dissipation from the heat sink to the atmosphere. Engine considered for the simulation parameters is of 120 cc/500 series. Hence, value of maximum velocity ($V_i$) is 75.66 m/s which is used as a limiting parameter to assign the inlet boundary conditions for the setup of numerical analysis [Figs. 1 and 2] [Table 1]

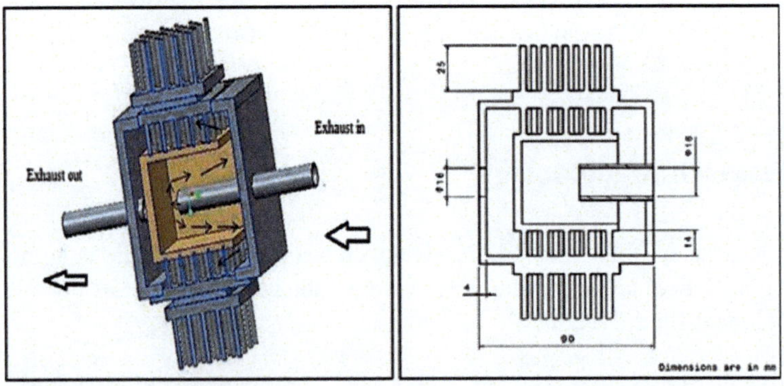


**Fig. 1** Sectional view of thermoelectric conversion device [19]

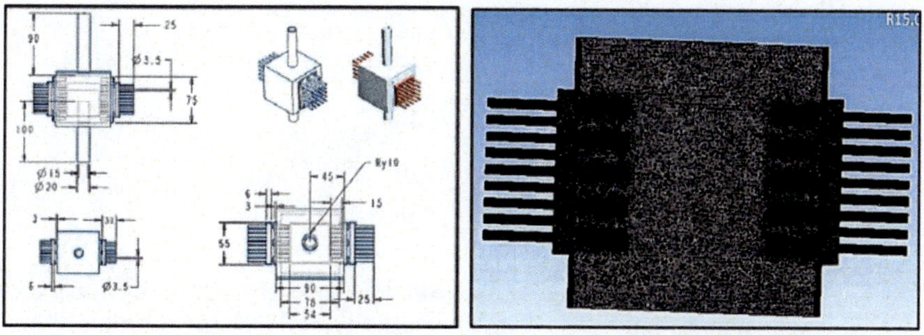


**Fig. 2** TEG conversion device and meshing [19]

**Table 1** Grid independence test

| S. No. | Nodes | Velocity variation (%) |
|---|---|---|
| 1 | 642 × 640 | – |
| 2 | 835 × 832 | 21.82 |
| 3 | 1090 × 1092 | 11.81 |
| 4 | 1200 × 1195 | 9.59 |
| 5 | 1400 × 1400 | 9.56 |

## 2.2 *Analytical Investigation*

The main purpose of doing the analytical study is to calculate the heat utilized which will be found by including the numerical data. Assuming heated air as our working fluid, properties of air are taken into consideration for the purpose of calculation.

### 2.2.1 Major Formulas to Be Incorporated in the Study

1. Performance of thermoelectric material

$$X = \alpha / Z_r \tag{1}$$

where $X$ is the figure of merit, $Z_r$ is the thermal conductivity, $\alpha$ is Seebeck coefficient, and $r$ is the electric resistivity.

2. Voltage output from thermoelectric generator

$$E = \Delta T \times \alpha \tag{2}$$

where $\Delta T$ is the temperature difference between source and sink.

3. Combined conductance of heat transfer and electrical resistance

$$k_s = \frac{Z_s A_s}{t_s} \tag{3}$$

$$R_s = \frac{r_s t_s}{A_s} \tag{4}$$

where $k_s$ is combined conductance of TEG single couple, $A_s$ is cross-sectional area of TEG single couple, and $r_s$ is electrical resistivity of TEG single couple.

4. Total heat transfer at the tip of fin

$$Q = \sqrt{P h_H Z_f A_f}(T_H - T_C)\frac{\tanh ml + \frac{h_H}{Z_f m}}{1 + \frac{h_H}{Z_f m}\tanh ml} \tag{5}$$

where

$$m = \sqrt{\frac{h_H C}{Z_f A_f}}$$

$h_H$ is convection heat transfer coefficient near hot section, $C$ is perimeter, $A_f$ is cross-sectional area of fin, $Z_f$ is conduction heat transfer coefficient of fin, $T_H$ is hot section temperature, and $T_C$ is cold section temperature.

5. Convective heat transfer coefficient for air

$$h_c = 10.45 - v + 10\sqrt{v} \tag{6}$$

where $v$ is relative speed between object surface and air (m/s).

To accomplish the purpose of analytical study, sections near thermoelectric generator are considered. Thermoelectric generator consisting of 126 p-n couples with dimension 40 × 40 × 3.8 mm has the following parameters.

- Thickness of the thermoelectric generator ($t$) = $3.8 \times 10^{-3}$ m
- Cross-sectional area of thermoelectric generator (AW) = $1.6 \times 10^{-4}$ $m^2$
- Area of single pair of thermoelectric generators (As) = $1.27 \times 10^{-5}$ $m^2$

To calculate the power production using the data acquired from simulation results, planes near the source and sink fins are made (Table 2).

**Table 2** Voltage, current and power output for TEGs

| S. No. | $\Delta T$ (K) | $E$ (V) | $I$ (A) | $P$ (W) |
|---|---|---|---|---|
| 1 | 10 | 0.5544 | 0.7253 | 0.4021 |
| 2 | 14 | 0.7762 | 1.029 | 0.7987 |
| 3 | 19 | 1.0534 | 1.397 | 1.4516 |
| 4 | 25 | 1.386 | 1.805 | 2.5 |

**Carnot Efficiency**

$$\eta_{\text{Carnot}} = 1 - \frac{Q_{\text{C}}}{Q_{\text{H}}} \tag{7}$$

Above formula is used in finding out the Carnot efficiency ($\eta_{\text{Carnot}}$) with calculated values of rate of heat supply or removal. This value with respect to convective heat transfer coefficient ($h_{\text{H}}$) is going to be a surprisingly interesting in explanation about the power output. Also, power loss ($P_{\text{loss}}$) estimation is also calculated.

## 3 Results and Discussion

The numerical analysis is performed on Ansys, which has provided us with temperature and velocity distribution on device. Inlet conditions: 50 m/s, 473.15 K [Figs. 3 and 4] [Table 3].

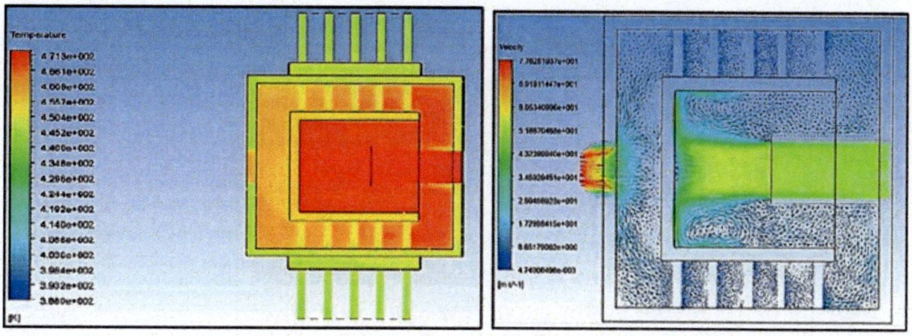

**Fig. 3** Numerical simulation representing temperature and velocity variations

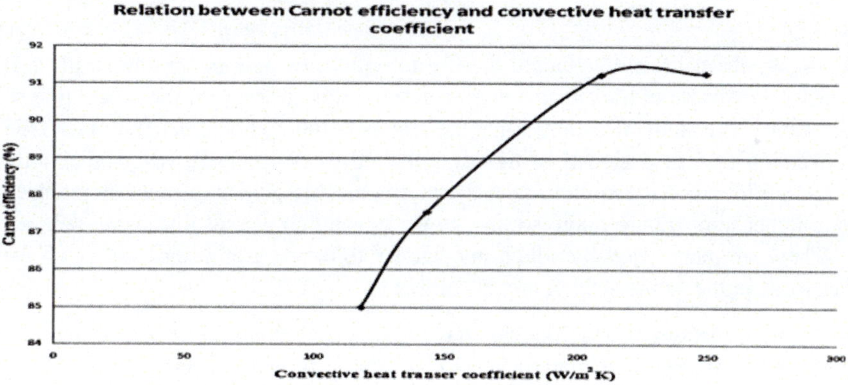


**Fig. 4** Carnot efficiency v/s convective heat transfer coefficient

**Table 3** Power, power loss and efficiencies of the thermoelectric device

| S. No. | 1 | 2 | 3 | 4 |
|---|---|---|---|---|
| Power available $P_{avail}$ (W) | 6.5 | 10.7 | 20 | 27.85 |
| Expected power output $P$ (W) | 0.4021 | 0.7987 | 1.4516 | 2.5 |
| Carnot efficiency $\eta_{Carnot}$ (%) | 85 | 87.56 | 91.24 | 91.28 |
| Power loss $P_{loss}$ (W) | 6.09 | 9.9 | 18.55 | 25.35 |
| Efficiency $\eta$ (%) | 5.26 | 6.53 | 6.68 | 8.19 |

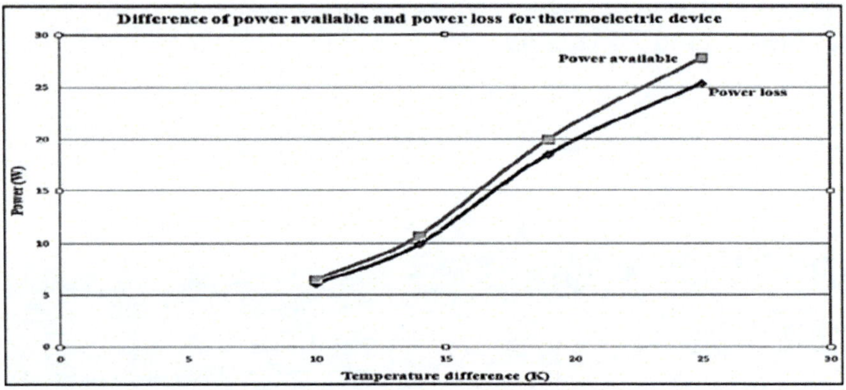


**Fig. 5** Power available and power loss versus temperature difference

Power loss from this device at minimum temperature difference is almost overlapping the available power curve (Fig. 5).

## 4 Conclusion and Future Scope

The research successfully studied the Carnot efficiency and its variation with heat transfer coefficient of the exhaust engine gases and used them for the production of electricity. The study also shows that the engine output is not affected by this TEG conversion system designed being used and, last but not least, if a higher temperature difference is required, then the thermoelectric generator must be replaced to a higher temperature range so that power generation can also be increased.

There are many ways by which one can try to improve efficiency on which we can work in the future, which are as follow:

- Optimum length at which device has to be placed along the exhaust pipe
- Materials of heat absorber and heat sink can be further studied.

## References

1. Olle H, Ronnie A (2016) A simulation framework for prediction of thermoelectric generator performance. Appl Energy 180:472–482
2. Yu J, Zhao H (2007) A numerical model for thermoelectric generator with parallel plate heat exchanger. J Power Sources 172:428–434
3. Bai S, Lu H, Wu T, Yin X, Shi X, Chen l (2014) Numerical and experimental analysis for exhaust heat exchangers in automobile thermoelectric generators. Case Stud Therm Eng 4:99–112
4. Montecucco A, Knox AR (2014) Accurate simulation of thermoelectric power generating systems. Appl Energy 118:166–172
5. Baskaya S, Karaaslan S, Calisir T, Zeki Yilmazoglu M, Yilmaz TO (2015) Experimental and numerical study on thermoelectric generator performance applied to a condensing combi boiler. Heat Transfer Eng 36(14–15):1292–1302
6. Lu C, Wang S, Chen C, Li Y (2015) Effects of heat enhancement for exhaust heat exchanger on the performance of thermoelectric generator. Appl Thermal Eng 89:270–279
7. Huang K, Yan Y, Li B, Li Y, Li K, Li J (2018) A novel design of thermoelectric generator for automotive waste heat recovery. Automot Innov 1:54–61
8. Kim KH, Patil MS, Seo JH, Kim C-J, Lee G-S, Lee M-Y (2018) Parametric study on heat transfer characteristics of waste heat recovery heat exchanger for automotive exhaust thermoelectric generator. Int J Eng Technol 7(2.33):6–10
9. Karri MA, Thacher EF, Helenbrook BT (2011) Exhaust energy conversion by thermoelectric generator: two case studies. Energy Convers Manag 52:1596–1611
10. Hsu C-T, Huang G-Y, Chu H-S, Yu B, Yao D-J (2011) Experiments and simulations on low temperature waste heat harvesting system by thermoelectric power generators. Appl Energy 88:1291–1297
11. Wu C (1996) Analysis of waste heat thermoelectric power generators. Appl Thermal Eng 16(1):63–69
12. Suzuki RO, Tanaka D (2003) Mathematical simulation on thermoelectric power generation with cylindrical multi-tubes. J Power Sources 124:293–298
13. Jadhav MG, Sidhu JS (2017) Design and fabrication of silencer waste heat power generation system using thermo-electric generator. 7(1):1–14
14. Kim TY, Negash A, Cho G (2017) Experimental and numerical study of waste heat recovery characteristics of direct contact thermoelectric generator. Energy Convers Manag 140:273–280
15. Yadav A, Pipe KP, Shtein M (2008) Fiber-based flexible thermoelectric power Generator. J Power Sources 175:909–913
16. Min G, Rowe DM (2007) Ring-structured thermoelectric module. Semicond Sci Technol 22:880–883
17. Jang J-Y, Tsai Y-C, Wu C-W (2013) A study of 3-D numerical simulation and comparison with experimental results on turbulent flow of venting flue gas using thermoelectric generator modules and plate fin heat sink. Energy 53:270e281
18. Jeng T-M, Tzeng S-C, Yang B-J, Li Y-C (2016) Design, manufacture and performance test of the thermoelectric generator system for waste heat recovery of engine exhaust. Inventions
19. Design, fabrication and performance analysis of automotive exhaust gas waste-heat recovery module using thermoelectric technology. Int J Emerg Technol Innov Res (www.jetir.org | UGC and issn Approved), ISSN:2349-5162, 5(12):970–983, December 2018, Available at : http://www.jetir.org/papers/JETIREB06114.pdf

# Experimental Investigation on the Performance of VCRs-Based Liquid Desiccant Dehumidification System Integrated with Cellulose Pads of Variable Flute Heights

**Minesh Vohra, Ankur Kalwar, Sanjeev Kumar, and Deepak Upadhyay**

**Abstract** The performance of liquid desiccant dehumidification system (LDDS) combined with vapor compression refrigeration system (VCRS) has been experimentally investigated. LDDS is used in order to enhance outdoor quality of air by extracting the moisture from humid air and VCRS deals with sensible cooling of air. The system performance has been enhanced by proper mixing of outdoor air with $CaCl_2$ as a desiccant in cellulose pads of different flute height in dehumidifier. The experiments have been executed at Lovely Professional University (LPU), Jalandhar, India [31°15′17.98″ (latitude) North and 75°42′56.382″ (longitude) East] at 55 blocks. In this study, moisture removal rate (MRR) and dehumidification effectiveness ($\varepsilon_{deh}$) are calculated to evaluate the performance of dehumidifier. Also, the coefficient of performance (COP) of VCRS with and without LDDS is calculated. These parameters are studied at different inlet air temperatures and humidity ratio. The maximum increase in $(\text{COP of VCRS})_{\text{with LDDS}}$ is 9.84%, when the temperature of air at the inlet is varying from 30 to 50 °C by keeping constant humidity ratio at inlet as 8 g/kg of dry air. The maximum increase in MRR is 51.58%, when inlet air humidity ratio is varying between 8 and 18.8 g per kg of dry air keeping constant inlet air temperature of 30 °C.

**Keywords** Liquid desiccant dehumidification system · Dehumidification · Regeneration · Humidity ratio · Liquid desiccant · Moisture removable rate (MRR)

M. Vohra (✉) · A. Kalwar · S. Kumar · D. Upadhyay
Lovely Professional University, Phagwara, India
e-mail: minesh.15783@lpu.co.in



R. Kumar et al. (eds.), *Recent Trends in Thermal Engineering*, Lecture Notes in Mechanical Engineering, https://doi.org/10.1007/978-981-16-3132-0_4

# 1 Introduction

As deficiency in energy seems an issue of rising concern in the universe, along with adverse effect of conventional refrigerants on the atmospheric conditions, hence requirement of emerging out with an alternative ecofriendly air conditioning system has become the point of major concern. In the past, Ertas et al. [1] had studied the properties of LiCl and $CaCl_2$ as a mixture of desiccant solution observed that the new inexpensive liquid desiccant comprising of 50:50 ratio of LiCl and $CaCl_2$ proved better in all aspects. Dai et al. [2] had investigated the performance aspects of vapors compression air conditioning by using liquid desiccant and concluded that the desired effect (Cooling) and COP of proposed hybrid system was increased significantly in comparison to VCS alone. Sultan et al. [3] studied the operational aspects of packed tower used in liquid desiccant regeneration and developed correlations to find out the rate at which evaporation occurs for water from the desiccant solution. Conde [4] explored physical properties of aqueous solutions along with enthalpy of dilution comprising of lithium and calcium chlorides and formulate its application in air conditioning equipment design. Zhang [5] fabricated the LiCl-based liquid membrane composite solution and analyzed the water vapor infusion and hence concluded that the moisture absorption capacity within membrane was around 1.14e−4 kg $m^{-2}$ $s^{-1}$ that is approximately double through solid hydrophilic cellulose acetate (SHCA) membrane with same dimensions. Liu et al. [6] observed experimentally mass and energy transfer in liquid and air-desiccant using a cross-flow regenerator with Li Br solution as a desiccant and conclude the effect of air with variation in inlet conditions of desiccant and correlation was proposed based on mass transfer. Hassan and Hassan [7] proposed a new liquid desiccant by mixing calcium nitrate using $CaCl_2$ in various proportions and calculated the thermophysical properties of LD. Mohan et al. [8] performed the depiction of column-based liquid desiccant using psychrometric correlations for hybrid air conditioner along with properties related to liquid desiccant. Zhang et al. [9] experimentally investigated the diffusion-based characteristics of a packed structured dehumidifier using a lithium chloride solution being a desiccant and reflects variation of air velocity with overall mass transfer coefficient. Zhang et al. [10] had done the analysis on advanced de-frosted air conditioner using LD dehumidification seasonal basis and concluded that to avoid frosting, specific humidity of air flowing through outer part of heat exchanger of VCHP is decreased, and COP was improved by 20% and 100% in winters and summers *b* in comparison to traditional air conditioning system. Bansal et al. [11] compared an internally cooled and an adiabatic cooled packed structured bed dehumidifier's performance using $CaCl_2$ in the cross-flow configuration and concluded that the maximum effectiveness and moisture removal were achieved with internal cooling. Bassuoni [12] done an experimental study on structured packing having a density 390 kg/$m^3$, corrugation angle-60° and void fraction 0.88 using $CaCl_2$ as a liquid desiccant. They concluded that the coefficient of mass transfer and dehumidification for each dehumidifier was found by elevating both air as well as solution's mass flow rate.

Yin et al. [13] carried out a performance analysis using hot air analytically with a validated mathematical model using desiccant solution regenerator and suggested that the maximum regeneration thermal efficiency was around 8. Jain et al. [14] investigated the LD system by incorporating a double pass heat exchanger for air to LD heat and mass transfer for reducing carryover. The result showed that LiCl performed better than $CaCl_2$ in dehumidification performance. Reference [15] carried out experiments on different hygroscopic membranes using LiCl in which a wicking material was sandwiched and concluded that polypropylene membrane showed vapors flux up to 1300 g/m$^3$. Huang et al. [16] investigated the quasi-counter flow parallel-plate membrane contactor (QFPMC) and concluded that the cooling and the dehumidification effectiveness of the QFPMC used for this system were reduced by approximately 5–29% and 2–13. Longo and Gaspar. Longo and Gasparella [17] studied experimentally the measurement of thermo-physical properties of desiccant comprises of $H_2O$/KCOOH (potassium formate) including concentration of salt from 60 to 80% in temperature range from 1 to 80 ° C. They concluded that the same desiccant stated above exhibits thermal conductivity of 23–33% lower than that of water and the dynamic viscosity 4–30 times higher than water.

Many researchers had experimentally investigated the liquid desiccant cooling system with different desiccants and combination of desiccants in dehumidifier. But no research work has been done on the hybrid packing with cellulose pads having different flute heights in dehumidifier. The performance of a liquid desiccant dehumidification system combined with vapor compression refrigeration system is presented. The objective of this manuscript is to improve the performance of absorption of moisture and the COP of system. The performance parameter such as moisture removal rate (MRR), dehumidification effectiveness ($\varepsilon_{deh}$), (COP of VCRS)$_{without\ LDDS}$, and (COP of VCRS)$_{(with\ LDDS)}$ has been investigated experimentally.

## 2 Experimental Setup

Experimental setup constitutes two sub-systems, namely liquid desiccant dehumidification system as well as vapor compression refrigeration system (VCRS). The liquid desiccant dehumidification system reduces humidity, and VCRS reduces the temperature for better cooling (Fig. 1).

The major components used in the experimental setup are dehumidifier, regenerator, cellulose pad, heater box, liquid desiccant, flat plate heat exchanger, and VCRS. The other components for performing the experiments are air blower, air heater, pump, duct, pipes, and axial fan. The flow of air is carried out in ducts, and flow of liquid desiccant is in pipes.

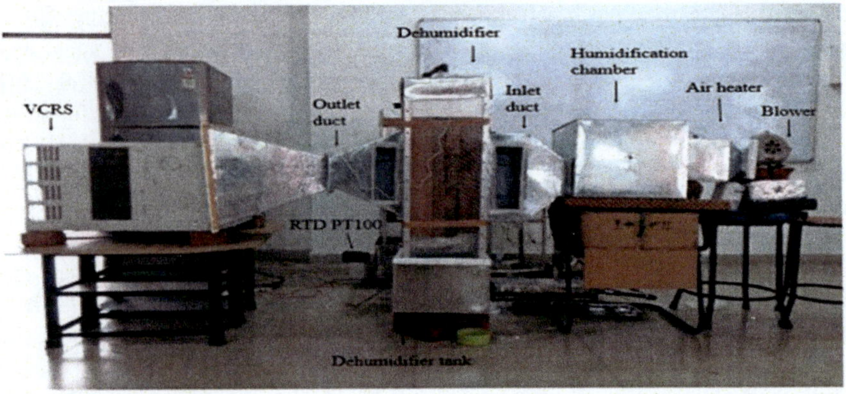


**Fig. 1** Experimental setup

# 3 Analysis of Experimental Data

The performance of a liquid desiccant dehumidification system integrated with vapor compression refrigeration system (VCRS) has been investigated based on moisture removal rate (MRR), dehumidification effectiveness ($\varepsilon_{deh}$), (COP of VCRS)$_{\text{without LDDS}}$, and (COP of VCRS)$_{\text{with LDDS}}$.

## *3.1 Moisture Removal Rate (MRR)*

Moisture removal rate is defined as the rate at which moisture is absorbed from inlet air and is calculated from Eq. (1) as

$$\text{MRR} = m_{\text{a,i}}\left(w_{\text{a,i}} - w_{\text{a,i}}\right) \tag{1}$$

where $w_{\text{a,i}}$ is absolute humidity of air at inlet; $w_{\text{a,o}}$ represents absolute humidity of outlet air, and $m_{\text{a,i}}$ is a mass flow rate of air at inlet.

Mass flow rate of the air at inlet is calculated from Eq. (2) as

$$m_{\text{a,i}} = \rho A v \tag{2}$$

where "$\rho$" is the density of the air; $A$: Area of the column and $v$ = Velocity of air.

The density of air varies with respect to pressure and temperature, so it is calculated by using ideal gas Eq. (3) as

$$\rho = \frac{P}{R_{\text{spec}} T} \tag{3}$$

where $p$ is the atmospheric pressure; $R_{\text{spec}}$ as specific gas constant, and $T$ is the temperature of air.

The density of air varies with the addition of water vapor in it, so the density of humid air is calculated by Eq. (4) as

$$\rho_{\text{humidair}} = \frac{P_{\text{d}} M_{\text{d}} + P_{\text{v}} M_{\text{v}}}{RT} \tag{4}$$

where $P_{\text{d}}$ is partial pressure of dry air; $P_{\text{v}}$: Partial pressure of water vapor; $M_{\text{d}}$: Molar mass of dry air; $M_{\text{v}}$: Molar mass of water vapor; with $R$ the universal gas constant and $T$: Temperature of air. Partial pressure of water vapor is calculated by Eq. (5) as

$$P_{\text{v}} = \varphi . P_{\text{sat}} \tag{5}$$

where $\varphi$ is the absolute humidity of the air and $P_{\text{sat}}$ is the partial pressure of saturated air.

Partial pressure of saturated air is calculated by Eq. (6) as

$$P_{\text{sat}} = 6.1078 \times 10^{\frac{7.5T}{T+237.3}} \tag{6}$$

Partial pressure of dry air is calculated by using Eq. (7) as

$$P_{\text{d}} = P - P_{\text{v}} \tag{7}$$

## 4 Results and Discussion

During experimentation, flow rate of $CaCl_2$ is maintained at 10 L/min at the inlet of dehumidifier, and air flow velocity at the inlet is maintained at 1.05 m/s. Three cases are formed to study system's performance:

4.1 Variation of MRR with temperature of air at inlet.

4.2 Absolute humidity of air at inlet varies by keeping constant inlet air temperature.

4.3 Comparison of COP of VCRS without LDDS and COP of VCRS with LDDS under different conditions.

## *4.1 Variation of MRR with Temperature of Air at Inlet*

Figure 2 shows the variation of MRR with respect to inlet air temperature. Inlet air temperature varies from 30 to 50 °C by keeping constant humidity ratio of inlet air of 8 g/kg of dry air. On increasing temperature of air, moisture removal rate of dehumidifier decreases due to lower difference of vapor pressure between the desiccant solution and the surrounding air. Thus, at higher temperature, moisture absorption capacity of desiccant decreases which results the lesser change in humidity ratio. So, the MRR decreases from 0.2294 to 0.0453 g/s when the inlet air temperature varies from 30 to 50 °C.

## *4.2 Absolute Humidity of Air at Inlet Varies by Keeping Constant Inlet Air Temperature*

The humidity ratio of inlet air is varying from 8 to 18.8 g per kg of dry air, and temperature of inlet air is maintained constant at 30 °C.

### 4.2.1 Variation of MRR with Absolute Humidity of Air at Inlet

Figure 3 shows the variation of MRR in dehumidifier with respect to absolute humidity of air at inlet. The absolute humidity of air at inlet is varying from 8 g/kg of dry air to 18.8 g/kg of dry air, and inlet air temperature is kept constant at 30 °C. The MRR increases on increasing air's absolute humidity, as higher the absolute humidity of air at inlet, the partial vapor pressure also increases leads to increase in pressure difference between the air at surrounding and the desiccant solution. Hence, MRR increases from 0.228 g/s to 0.346 g/s when the absolute humidity of air at inlet varies from 8 to 18.8 g/kg of dry air.

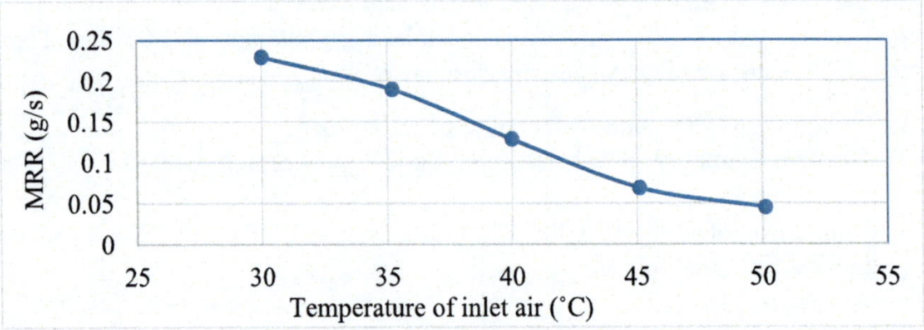


**Fig. 2** Variation of MRR w.r.t temperature of air at inlet

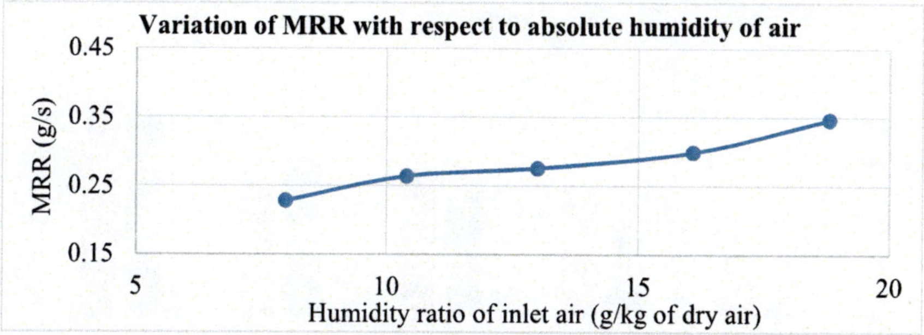


**Fig. 3** Variation of MRR w.r.t. absolute humidity of air at inlet

## *4.3 Comparison of COP of VCRS Without LDDS and COP of VCRS With LDDS Under Different Conditions*

### 4.3.1 Comparison of COP of VCRS With and Without LDDS at Variable Temperature of Air at the Inlet

Figure 4 presents the comparison of $(\text{COP of VCRS})_{\text{without LDDS}}$ with $(\text{COP of VCRS})_{\text{with LDDS}}$ with respect to temperature at the inlet conditions of air. The temperature of air at the inlet varies from 30 to 50 °C by keeping constant inlet absolute humidity of air as 8 g/kg of dry air. It is clear from Fig. 4 that the $(\text{COP of VCRS})_{\text{with LDDS}}$ increases more as compare to the $(\text{COP of VCRS})_{\text{without LDDS}}$ with rise in air temperature at the inlet. Due to incremented temperature of air at the inlet conditions, sensible load of air increases which results in the increment of enthalpy difference of air, covered mainly by VCRS.

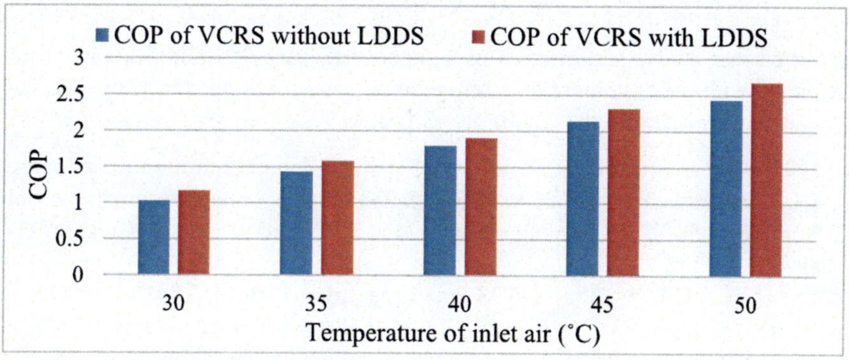


**Fig. 4** Comparison of $(\text{COP of VCRS})_{\text{without LDDS}}$ and $(\text{COP of VCRS})_{\text{with LDDS}}$ w.r.t. inlet air temperature

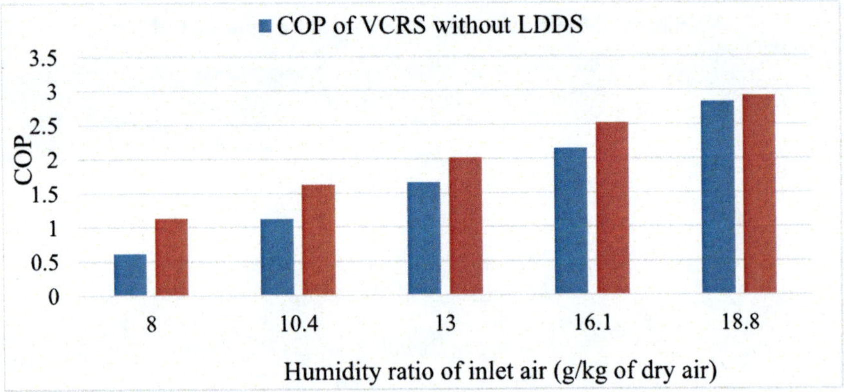


**Fig. 5** Comparison of (COP of VCRS)$_{\text{without LDDS}}$ and (COP of VCRS)$_{\text{with LDDS}}$ w.r.t. inlet air humidity ratio

### 4.3.2 Comparison of COP of VCRS with and Without LDDS at Different Inlet Absolute Humidity of Air

Figure 5 shows the comparison of (COP of VCRS)$_{\text{without LDDS}}$ and (COP of VCRS)$_{\text{with LDDS}}$ with respect to inlet air absolute humidity. The absolute humidity of inlet air is varying from 8 g to 18.8 g/kg of dry air, and inlet air temperature is kept constant at 30 °C. The figure shows that the (COP of VCRS)$_{\text{with LDDS}}$ increases more as compare to the (COP of VCRS)$_{\text{without LDDS}}$ as the absolute humidity of the air increases. Higher the latent load, higher will be the COP. Hence, the (COP of VCRS)$_{\text{with LDDS}}$ increases more as compared to the (COP of VCRS)$_{\text{without LDDS}}$.

## 5 Conclusion

In the current study, performance of a liquid desiccant dehumidification system integrated with vapor compression refrigeration system has been investigated. The conclusions drawn from the results are as follows:

1. MRR decreases from 0.2294 to 0.0453 g/s and dehumidification effectiveness ($\varepsilon_{deh}$) decreases from 0.1629 to 0.0343, when the inlet air temperature is varying from 30 to 50 °C by keeping constant inlet air humidity ratio of 8 g/kg of dry air.
2. C.O.P of VCRs without LDDS increases from 1.03 to 2.44 and (COP of VCRS)$_{\text{with LDDS}}$ increases from 1.17 to 2.68, when inlet air temperature is varying from 30 to 50 °C by keeping constant inlet air humidity ratio of 8 g/kg of dry air.

3. The (COP of VCRS)$_{without\ LDDS}$ increases from 0.62 to 2.83 and (COP of VCRS)$_{with\ LDDS}$ increases from 1.14 to 2.92, when at the inlet the specific humidity is varying from 8 to 18.8 g/kg of dry air by keeping constant inlet air temperature of 30 °C.
4. The proposed hybrid system is effective when latent heat ratio of air is higher than sensible heat ratio.
5. Hybrid packing of cellulose pad in dehumidifier results in better performance in all the aspects.

## References

1. Ertas A, Anderson E, Kiris I (1992) Properties of a new liquid desiccant solution—lithium chloride and calcium chloride mixture. Solar Energy 49:205–212
2. Dai YJ, Wang RZ, Zhang HF, Yu JD (2001) Use of liquid desiccant cooling to improve the performance of vapour compression air conditioning. Appl Therm Eng 21:1185–1202
3. Sultan GI, Hamed AM, Sultan AA (2002) The effect of inlet parameters on the performance of packed tower-regenerator. Renew Energy 26:271–283
4. Conde MR (2004) Properties of aqueous solutions of lithium and calcium chlorides: formulations for use in air conditioning equipment design. Int J Therm Sci 43:367–382
5. Zhang L-Z (2006) Fabrication of a lithium chloride solution based composite supported liquid membrane and its moisture permeation analysis. J Membr Sci 276:91–100
6. Liu XH, Jiang Y, Chang XM, Yi XQ (2007) Experimental investigation of the heat and mass transfer between air and liquid desiccant in a cross-flow regenerator. Renew Energy 32:1623–1636
7. Hassan AAM, Hassan MS (2008) Dehumidification of air with a newly suggested liquid desiccant. Renew Energy 33:1989–1997
8. Mohan S, Maiya MP, Tiwari S (2008) Performance characterization of liquid desiccant columns for a hybrid air-conditioner. Appl Therm Eng 28:1342–1355
9. Zhang L, Hihara E, Matsuoka F, Dang C (2010) Experimental analysis of mass transfer in adiabatic structured packing dehumidifier/regenerator with liquid desiccant. Int J Heat Mass Transfer 53:2856–2863
10. Zhang L, Dang C, Hihara E (2010) Performance analysis of a no-frost hybrid air conditioning system with integrated liquid desiccant dehumidification. Int J Refrig 33:116–124
11. Bansal P, Jain S, Moon C (2011) Performance comparison of an adiabatic and an internally cooled structured packed-bed dehumidifier. Appl Therm Eng 31:14–19
12. Bassuoni MM (2011) An experimental study of structured packing dehumidifier/regenerator operating with liquid Desiccant. Energy 36:2628–2638
13. Yin Y, Li S, Zhang X, Peng D (2011) Feasibility and performance analysis of a desiccant solution regenerator using hot air. Energy Build 43:1097–1104
14. Jain S, Tripathi S, Das RS (2011) Experimental performance of a liquid desiccant dehumidification system under tropical climates. Energy Convers Manag 52:2461–2466
15. Das RS, Jain S (2013) Experimental performance of indirect air liquid membrane contactors for liquid desiccant cooling systems. Energy 57:319–325
16. Huang S-M, Yang M, Yang X (2014) Performance analysis of a quasi-counter flow parallel-plate membrane contactor used for liquid desiccant air dehumidification. Appl Therm Eng 63:323–332
17. Longo GA, Gasparella A (2016) Experimental measurement of thermo physical properties of H2O/KCOOH (potassium formate) desiccant. Int J Refrig 62:106–113

# Optimization of Aerofoil Profiles Using a Notch on a Vertical Axis Wind Turbine

**Amit Kumar Thakur and Ajay Kumar Kaviti**

**Abstract** This paper focuses on the principle and results of the alteration of the straight wind vertical axis wind turbine. A modification to the aerofoil profile is tested using a notch. This research focuses on the lift coefficient values, the drag, and moment coefficient as these are the essential parameters used to calculate the turbine power provided by inviscid turbulence models which are used to simulate the mean flow characteristics of turbulent flow in computational fluid dynamics. Both design modifications and simulation research were done using ANSYS Fluent, utilizing computational fluid dynamics. In ANSYS, the FLUENT (CFX) module is used for performing the simulation process. Standard profile blades are analyzed and are compared with that of the modified profile blade values. In this analysis process, symmetrical 4-digit aerofoil models such as NACA0012, NACA0015, and NACA0018 are used. The whole analysis was conducted based on the lift and drag characteristics of each profile at 2°, 4°, 8°, 10° angles and compared these with the standard ones. Inviscid flow models are used to avoid unnecessary friction for the flow, and the flow is presumed to be ideal flow to prevent the backflow pressure. This fundamental research is based on how the forces act on the aerofoil profile when fluid strikes with different velocities and at a different angle of attacks.

**Keywords** VAWT · ANSYS · NACA aerofoils · Notch

A. K. Thakur (✉)
Department of Mechanical Engineering, Lovely Professional University, Phagwara, Punjab, India
e-mail: amit.25010@lpu.co.in

A. K. Kaviti
Department of Mechanical Engineering, VNRVJIET, Hyderabad, India



R. Kumar et al. (eds.), *Recent Trends in Thermal Engineering*, Lecture Notes in Mechanical Engineering, https://doi.org/10.1007/978-981-16-3132-0_5

## 1 Introduction

Energy generated typically from natural resources is known as renewable energy, and consumption of these is exceeding day by day. Talking about wind power, there are two types of wind turbines, vertical, and horizontal axis wind turbines. HAWT is mostly used for large-scale power generation about the range from 1 kW to 10 MW. At the same time, the VAWT is available in 10 MW. Wind turbines rely on the blades to produce energy, so, the shape of the aerofoil is essential and directly affects the use of wind energy of the wind turbine. Advantage of VAWT is that it works efficiently at low-velocity wind, while HAWT requires wind with more velocity. Hence, VAWT can be used in any area, with enough wind so, VAWT will be playing a significant role in the future in the wind energy sector. Shukla et al. [1] explained the necessary information of the aerofoil profiles and the outline of the NACA 4-digit aerofoil analysis. They examined the effects of profile changes on straight bladed VAWTs, with NACA0012 and NACA0015 symmetric aerofoil profiles, which showed a higher coefficient of lift, velocity, and force. Jin et al. [2] studied the critical basic approaches and their implementations in Darrieus VAWTs, intending to have more experts know the current state of the work and also offering guidelines for a related study. Kumar et al. [3] highlighted some of VAWT's major developments, with an emphasis on urban infrastructure integration. Rezaeiha et al. [4] advocated the optimum aerodynamic design of small to large VAWTs and explored the effect; these parameters will have on the aerodynamic efficiency to provide a deeper insight into their implications. Li et al. [5] explained about the aerodynamic stability of air foils in lift-type vertical axis wind turbine in steady solver and examined static forces and moment of air foil in steady solver. Ghasemian et al. [6] analyzed recently published studies on Darrieus VAWT's computational fluid dynamics (CFD) simulations to discuss numerous developments in simulation fields. Ma et al. [7] have offered an integral approach to the question of developing different airfoil profiles that directly influence the capture ratio of wind power. The main goal of this work was to build an automated optimization program for airfoil profiles to boost a VAWT's power output. Ruiz et al. [8] examined how a traditional Savonius blade enhanced its performance by adding a new design without costly and complicated geometry adjustments. The two-dimensional CFD simulations have been concluded by Bianchini et al. [9] to provide reasonable computation costs and accurate estimates of turbine performance, as well as the assisted flow field around the rotor and its wake. Biber Stall [10] studied the slotted flap airfoil with focus on hysteresis. Hao et al. [11] evaluated the deformable flaps under active control conditions. Zhu et al. [12, 13] performed a simulation analysis of different types of airfoils.

This research tries to investigate the effect of profile modifications by energy and the inviscid models on straight bladed VAWTs equipped with symmetric aerofoil (NACA 4-digit NACA0012, NACA0015, and NACA0018 series). Alterations of the aerofoil profile are studied using a notch at various angles of flaps. In terms of lift coefficient (CL), drag, and moment, the effects of the changed profiles are compared. These findings are further checked under the same boundary conditions with previous experimental results to ensure analytical reliability.

## 2 Methodology

### *2.1 Study of Domain*

Aerofoil investigations with the help of analytical tools (ANSYS Fluent) will be conducted here. A domain with an aerofoil reveals enough areas of the domain of an aerofoil in the middle of this analysis to assess the effect of wind on the airfoil. For airfoil, the chord length here is "1 m," so a rectangular area domain and 12.5 m radius semi-circle is considered. Figure 1 depicts the aerofoil Profile for vertical axis wind turbine (VAWT) blades. With the support of the aerofoil at the center, an area around the aerofoil is constructed around which limiting conditions are applied. The surface-based design without thickness was considered to have consistent results for better effects for the whole domain and the aerofoil. The mesh variety, such as the external domains, network mesh control, and aerofoil (wall), is considered in different parts of a network. During this analysis, the mesh with 5235 nodes and 51,846 elements in its size is considered to be active in the domain. The final mesh is rendered when the proximity and curvature are determined, as our profile contains all geometry. The body sizing method is used for separating and refining the mesh control domain. The mesh control domain has a better mesh than the other domain region for better aerofoil performance. Here the unit scale of 500 mm with a growth rate of 1.20 is used. The blade edge method was used to distinguish this aerofoil mesh region from another region, with 100 division numbers on curved section and 150 division numbers on aerofoil end.

The limitations and methods used in the development of meshes are discussed in this article. In the three aerofoils, the above techniques and conditions are used continuously for creating a mesh.

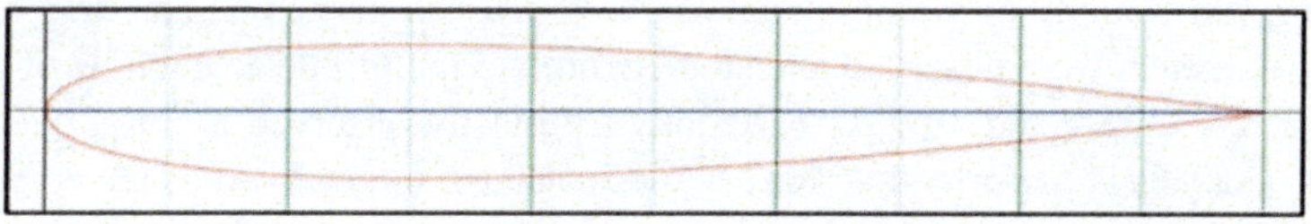

**Fig. 1** Aerofoil profile for vertical axis wind turbine (VAWT) blades

## 2.2 Simulation Analysis

*Computational Simulation*

All the blades are built, simulated, and evaluated with the assistance of ANSYS Fluent program. Required external domain region is considered for the function to be performed. An inner blade is filled with a circular domain for mesh refinement. The circular mesh is refined from the main external domain, and the more acceptable blade profile range is regarded as better results and as a way to avoid the impact of the outlet cap. The input parameters and the values we used in our simulation are indicated on the constant chord length C 1⁄4 1 m during service. The longitude of all chords is the same. The form and configuration of the aerofoil are adjusted, and the input parameters and boundary conditions for both primary and modified type aerofoils are kept constant. The values of the various blades (NACA0012, NACA0015, and NACA0018) are evaluated through this study. For both analyses, the boundary conditions and the domain field remain unchanged. We are based on one constant value of the Reynolds (Re) here. Simple blade analysis is done, focusing on the results of lift, velocity and force.

# 3 Results and Discussion

The profiles considered are NACA0012, NACA0015, and NACA0018, which are symmetrical concerning their chords. Aerofoil modifications include bulge, notch, and spine. In the present research, the modification to all aerofoil profiles is done using a notch to improve and observe the aerodynamic characteristics, which are analyzed in ANSYS workbench software and plane flap at angles 2°, 4°, 6°, 8°, 10°, 12°, and 15° using XFLR5 software which uses direct analysis.

## 3.1 NACA0012 with a Notch

The NACA0012 is modified to improve and observe the aerodynamic characteristics. The graphs of pressure–velocity contours are shown in Figs. 2 and 3. As pressure and velocity play a dominant role in the variations of lifts and drag forces, these parameters were observed and studied, around the profile at an angle of attack of air at 2°, 4°, 8°, and 10°. As variations are mostly observed at initial and final angles, analysis is performed at these angles of attack of air. Most of the eddies are observed to be prominent at tailing edge, so the profile near the trailing edge is modified. As the angle of attack increases, an increase in the pressure is observed due to more surface area is affected as the angle of attack increases. The pressure

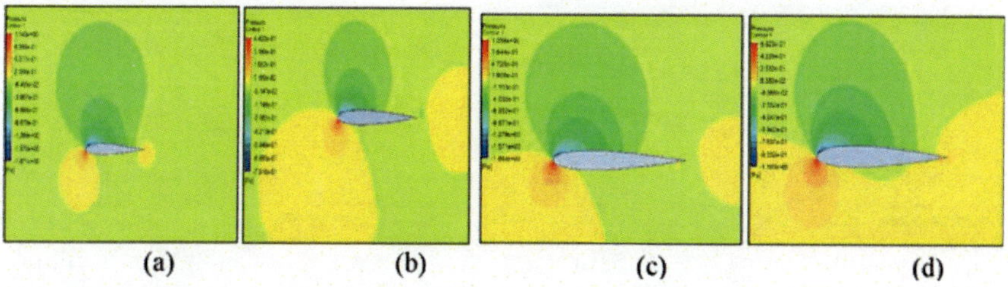

**Fig. 2** Pressure contour for NACA0012 with a NOTCH at an angle of attack **a** 2° **b** 4° **c** 8° **d** 10°

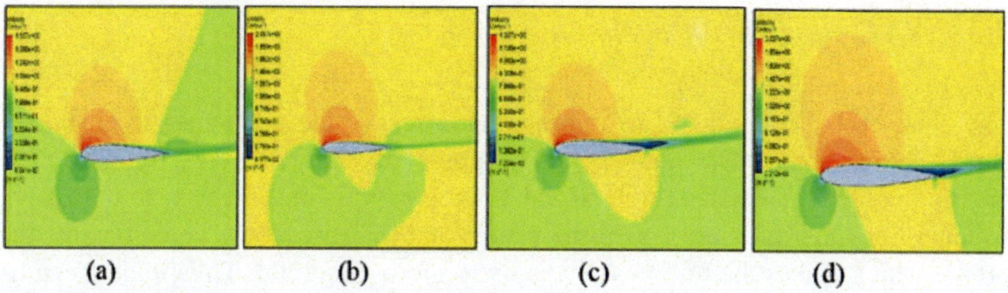

**Fig. 3** Velocity contour for NACA0012 with a NOTCH at an angle of attack **a** 2° **b** 4° **c** 8° **d** 10°

contour under the curve at the leading edge is observed to be maximum which allows low pressure above the profile and increase in the velocity over the profile which makes the turbine to have more lift to drag ratio. It can be observed through velocity contours of the profile at 2°, 4°, 8°, 10° angles of attack of air.

The notch is provided at the trailing edge of the profile, i.e., at 90% of the length of the cord. Practically, providing a notch would reduce the strength of the blade at the trailing edge. Due to this notch near tailing edge, the pressure distribution is disturbed. As the flow of air reaches the notch, the flow gets deviated from it and results in more pressure at the notch region. The notch region has more pressure as observed. Due to this, internal stresses will develop in that region and hence can be affected to wear and tear.

The stresses increase with an increase in the angle of attack. The single notch abducts the pressure distribution; the values of lift to drag ratio get decreased, and hence, it is not suggested for use. To avoid this, more number of notches must be implemented so that stress concentration at a single notch can be avoided.

The cl/cd values of NACA0012 (notch) constantly increase with the increase in the angle of attack, and then gradually fall at an angle of 10°. This phenomenon is because as the thickness of the profile is less and has less surface area. At 10° most of the air flows under the surface and hence significantly less amount of airflow is over the surface as shown in Fig. 10 so, at 10° angle of attack, the cl/cd value drops.

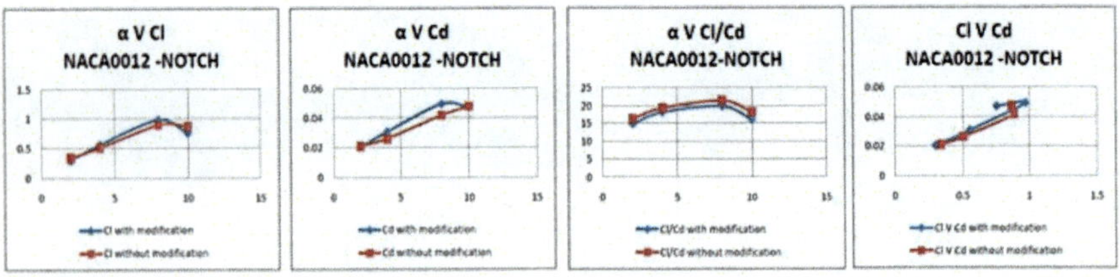


(a) (b) (c) (d)

**Fig. 4** Variation of various parameters for NACA0012 with a NOTCH at an angle of attack **a** 2° **b** 4° **c** 8° **d** 10°

## *3.2 NACA0015 with a Notch*

The NACA0015 is modified to improve and observe the aerodynamic characteristics. The graphs of pressure–velocity contours are shown in Figs. 5 and 6. The pressure contour under the curve at the leading edge is observed to be maximum which allows low pressure above the profile and increase in the velocity over the profile which makes the turbine to have more lift to drag ratio. This process can be observed through velocity contours of the profile at 2°, 4°, 8°, 10° angles of attack of air.

The notch is provided at the trailing edge of the profile, i.e., at 90% of the length of the cord. The pressure contour at the notch region can be observed. The notch region has more pressure. Due to this, internal stresses will develop in that region and hence can be affected to wear and tear.

The stresses increase with the increase in the angle of attack. Because there is more flow of air toward the notch due to the increase in the angle of attack. The single notch abducts the pressure distribution; the values of lift to drag ratio get decreased, and hence, it is not suggested for use. To avoid this, more number of notches must be implemented so that stress concentration at a single notch can be avoided.

The Cl/Cd values of NACA0015 continuously increased concerning the angle of attack and then gradually fall at an angle of 8° as depicted in Fig. 7. This behavior is due to more pressure is built up under the surface. The pressure contour shows that the pressure is acting over the leading edge toward the trailing edge. This

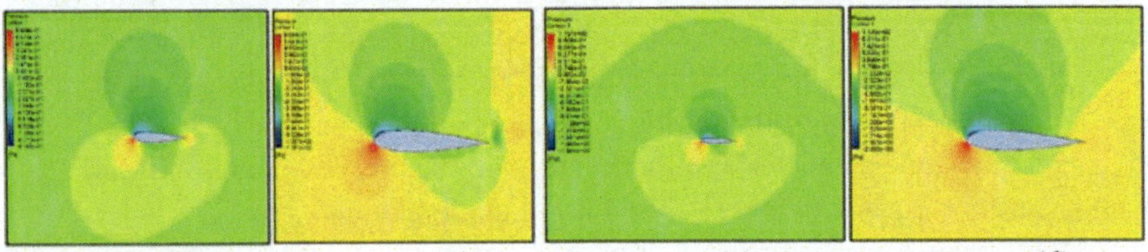

(a) (b) (c) (d)

**Fig. 5** Pressure contour for NACA0015 with a NOTCH at an angle of attack **a** 2° **b** 4° **c** 8° **d** 10°

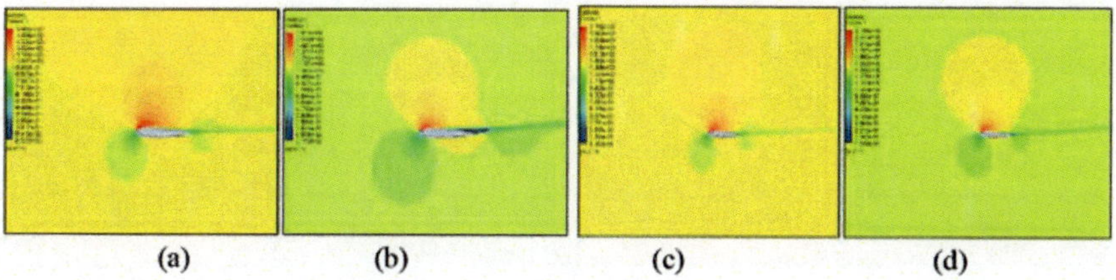


**Fig. 6** Velocity contour for NACA0015 with a NOTCH at an angle of attack **a** 2° **b** 4° **c** 8° **d** 10°

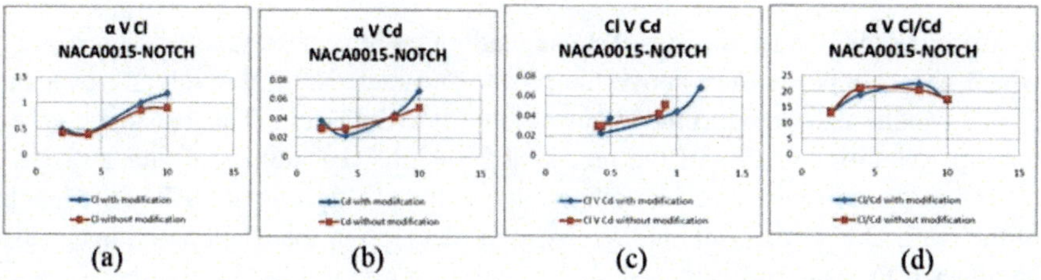


**Fig. 7** Variation of various parameters for NACA0015 with a NOTCH at an angle of attack **a** 2° **b** 4° **c** 8° **d** 10°

results in the highest Cl/Cd value at 8° angle of attack of air. As the angle increases, the velocity of air increases over the surface and leads to create low pressure over the surface, and hence, the pressure under the surface increases at the notch region and thus results in a decrease in Cl/Cd values after 8° angle of attack of air.

## 3.3 NACA0018 with a Notch

The NACA0018 is modified to improve and observe the aerodynamic characteristics. The graphs and pressure–velocity contours are shown in Figs. 8 and 9. The

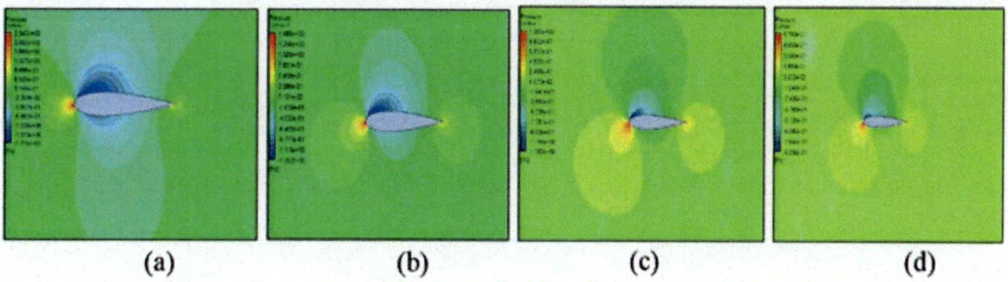


**Fig. 8** Pressure contour for NACA0018 with a NOTCH at an angle of attack **a** 2° **b** 4° **c** 8° **d** 10°

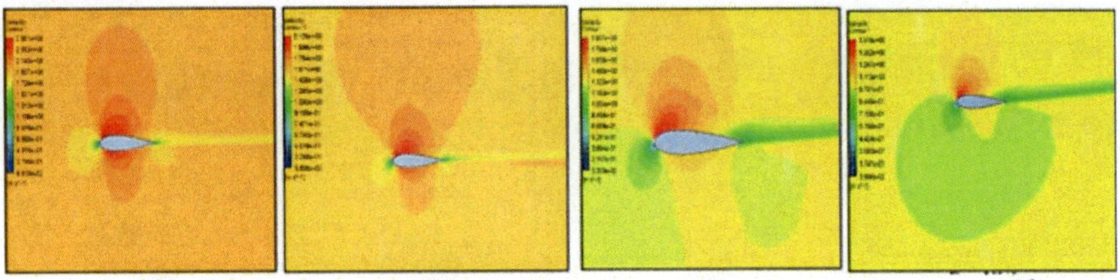

(a) (b) (c) (d)

**Fig. 9** Velocity contour for NACA0018 with a NOTCH at an angle of attack **a** 2° **b** 4° **c** 8° **d** 10°

pressure contour under the curve at the leading edge is observed to be maximum which allows low pressure above the profile and increase in the velocity over the profile which makes the turbine to have more lift to drag ratio. This can be observed through velocity contours of the profile at 2°, 4°, 8°, 10° angles of attack of air.

The notch is provided at the trailing edge of the profile, i.e., at 90% of the length of the cord. Due to this, internal stresses will develop in that region and hence can be affected to wear and tear.

The stresses increase with the increase in the angle of attack. Because there is more flow of air toward the notch due to the increase in the angle of attack. The single notch abducts the pressure distribution; the values of lift to drag ratio get decreased, and hence, it is not suggested for use. To avoid this, more number of notches must be implemented so that stress concentration at a single notch can be avoided.

The cl/cd values of NACA0018 (notch) constantly increase with the increase in the angle of attack and then gradually fall at an angle of 10°. This is because as the thickness of the profile is less and has less surface area. At 10° most of the air flows under the surface and hence significantly less amount of airflow is over the surface as shown in Fig. 10 so, at 10° angle of attack, the cl/cd value drops.

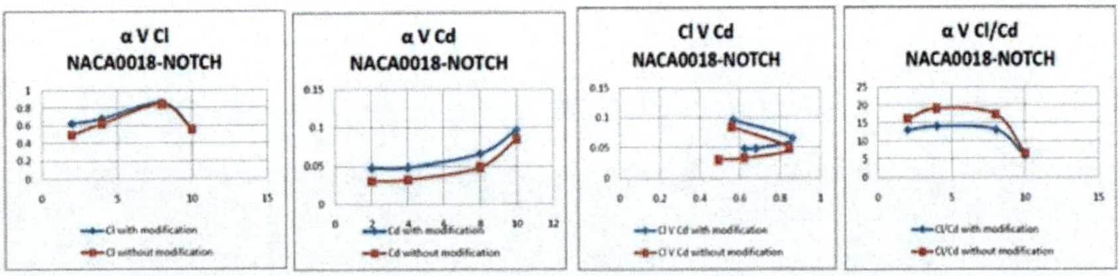


(a) (b) (c) (d)

**Fig. 10** Variation of various parameters for NACA0018 with a NOTCH at an angle of attack **a** 2° **b** 4° **c** 8° **d** 10°

## 4 Conclusions

Compared to the low angle of attack, the lift and drag coefficients of the aerofoils have a low value in lift/drag ratio for NACA0018 and a high value for NACA0012. Hence, NACA0012 is ideally suited for improvement of performance and output. However, NACA0012 is considered too thin (12% of chord length). Therefore, NACA0015 is chosen for aerodynamic stability and optimum results. In the practical analysis, when the thickness of the profile is decreased, the weight and centrifugal forces acting over it will reduce the strength of the blade. Therefore, for the best-optimized results, the distortion in shape should be considered when selecting the blade. All commercial uses of NACA 4-digit series were abandoned 70–80 years ago. Hence, the reason is the low lift to drag characteristics. As years passed, the need for high output with low inputs has increased. Design modifications will be prominent for such expected results.

The most used aerofoils are NACA0012, NACA0015, and NACA0018, keeping in mind the values of lift but not Cl/Cd ratio. Notch used in the aerofoil is one of the most substantial improvements that have been made over the last few projects. The raise (CL) reached by notch aerofoil is smaller than the primary blade for all forms of aerofoils we used. The maximum value Cl/Cd noted in aerofoil NACA0015 is 22.58 at 8° angle of attack. From all the flaps and profiles at various angles, only at 15° flap angle has highest Cl/Cd value for every profile. From overall modification analysis, we conclude that NACA0015 is an optimum thick profile for efficient performance of the VAWT.

## References

1. Shukla V, Kaviti AK (2017) Performance evaluation of profile modifications on straight-bladed vertical axis wind turbine by energy and Spalart Allmaras models. Energy 126:766–795
2. Jin X, Zhao G, Gao K, Ju W (2015) Darrieus vertical axis wind turbine: Basic research methods. Renew Sustain Energy Rev 42:212–225
3. Kumar R, Raahemifar K, Fung AS (2018) A critical review of vertical axis wind turbines for urban applications. Renew Sustain Energy Rev 89:281–291
4. Rezaeiha A, Montazeri H, Blocken B (2018) Characterization of the aerodynamic performance of vertical axis wind turbines: impact of operational parameters. Energy Convers Manage 169:45–77
5. Li J, Cao Y, Wu G, Miao Z, Qi J (2017) Aerodynamic stability of airfoils in a lift-type vertical axis wind turbine in steady solver. Renewable Energy 111:676–687
6. Ghasemian M, Ashrafi ZN, Sedaghat A (2017) A review on computational fluid dynamic simulation techniques for Darrieus vertical axis wind turbines. Energy Convers Manage 149:87–100
7. Ma N, Lei H, Han Z, Zhou D, Bao Y, Zhang K, Chen C (2018) Airfoil optimization to improve power performance of a high-solidity vertical axis wind turbine at a moderate tip speed ratio. Energy 150:236–252

8. Ostos I, Ruiz I, Gajic M, Gómez W, Bonilla A, Collazos C (2019) A modified novel blade configuration proposal for a more efficient VAWT using CFD tools. Energy Convers Manage 180:733–746
9. Bianchini A, Balduzzi F, Bachant P, Ferrara G, Ferrari L (2017) Effectiveness of two-dimensional CFD simulations for Darrieus VAWTs: a combined numerical and experimental assessment. Energy Convers Manage 136:318–328
10. Biber Stall K (2015) Hysteresis of an airfoil with slotted flap. J Aircraft 42(6):1462–1470
11. Hao W, Li C, Ye Z (2017) Computational study on wind turbine airfoils based on active control for deformable flaps. J Mech Sci Technol 31:565–576
12. Zhu H, Hao W, Li C (2018) Simulation on flow control strategy of synthetic jet in an vertical axis wind turbine. Aero Sci Technol 77:439–448
13. Zhu H, Hao W, Li C (2019) Numerical investigation on the effects of different wind directions, solidity, airfoils and building configurations on the aerodynamic performance of building augmented vertical axis wind turbines. Int J Green Energy 16:1681427

# Numerical Simulation of the Turbulent Natural Convection in a Square Enclosure Filled with Water Based $Al_2O_3$ Nanofluids

**Anuj Kumar Mishra, Ajay Kumar, Himanshu Tripathi, Naveen Sharma, Sumit Kanchan, and Rajesh Choudhary**

**Abstract** A CFD simulation of the buoyancy—induced convection in a square enclosure filled with distilled water (DW) based $Al_2O_3$ nanofluids for Rayleigh number (*Ra*) range of $10^9$–$10^{12}$ is conducted. The enclosure is heated from below by applying constant heat flux condition and cooled from above by the constant temperature boundary condition, with the adiabatic condition at walls, which makes the test section a Rayleigh-Bénard cavity. The particle size of 40 nm and particle concentration of 0.01 and 0.1 vol. % is used for $Al_2O_3$ nanoparticles. Results of the numerical work shown enhanced heat transfer at low concentration (0.01 vol. %) with the increase in *Ra*, while the opposite trend is observed for higher concentration (0.1 vol. %). The velocity of the fluid particles increased with the *Ra* and increased the turbulence in the test section. The eddy currents formed near the corners of the test section increased the resistance to the heat transfer, and the resistance further become strong with the *Ra*. A deterioration in the heat transfer occurred due to increased viscosity at higher *Ra* for both the concentrations of nanoparticles.

**Keywords** Nanofluid · Convection · Numerical simulation · Nusselt number · Rayleigh number · Rayleigh-Bénard cavity · Turbulent

---

A. K. Mishra
Department of Mechanical Engineering, Meerut Institute of Engineering and Technology, Meerut, Uttar Pradesh, India

A. Kumar · H. Tripathi
Department of Mechanical Engineering, Laxmi Devi Institute of Engineering and Technology, Alwar, Rajasthan, India

N. Sharma
Department of Mechanical Engineering, DVR & Dr. HS MIC College of Technology, Kanchikacherla, Andhra Prades, India

S. Kanchan · R. Choudhary (✉)
School of Mechanical Engineering, Lovely Professional University, Phagwara, Punjab, India
e-mail: dr.rajesh1316@gmail.com

 
R. Kumar et al. (eds.), *Recent Trends in Thermal Engineering*, Lecture Notes in Mechanical Engineering, https://doi.org/10.1007/978-981-16-3132-0_6

# 1 Introduction

The nanosized solid particles can alter the characteristics of the fluid flow and heat transfer. In 1995, Choi and Eastman [1] used the suspension of nanosized solid particles to overcome the existing problems with micron-sized solids. The dispersion of solid particles (diameter = 1–100 nm) uniformly in the conventional fluid, defined as nanofluids, leaded to increased thermal conductivity in nanofluids compared to the micron sized particles due to higher Brownian motion of nanosized particles [2–5]. Since, in the buoyancy—induced convection heat transfer, no external force involved to induce the velocity and it is only occurred due to buoyancy forces, which caused to a lower heat transfer capability. Therefore, to improve the natural convection, effect of particles suspension on the heat transfer capability of the conventional working fluids have been studied using both the approaches, numerically and experimentally.

Hwang et al. [6] reported that the effective heat transfer decreased with the nanoparticle's diameter, and with the decrement in the average temperature of nanofluids. Another study conducted by the Oztop and Abu-Nada [7] to study the different nanofluids in a cavity heated partially and experienced that the Rayleigh number and the particle loading is primary functions to improve the rate of heat transfer in the buoyancy induced heat transfer. Cianfrini et al. [8] used the SIMPLE-C algorithm to simulate the buoyancy—induced heat transfer in water based $Al_2O_3$ nanofluids filled in a cavity which is heated partially from the side. An increased heat transfer with the particle loading was observed and also found that by increasing the diameter of nanoparticles, width of cavity or aspect ratio and length of heater, the heat transfer enhancement due to suspension of nanoparticles and optimal loading of particle concentration were decreased. Abouali and Ahmadi [9] presented a review on the computational analysis of the natural convection in the different geometries, like Rayleigh-Bénard cavity, triangular enclosure, tilted horizontal squares, vertical and horizontal annulus, etc., filled with the CuO and $Al_2O_3$ nanofluids. The authors reported an improvement in the heat transfer coefficient with the concentration of nanoparticles up to an optimal value and beyond, heat transfer reduced. The optimal value of the particle concentration was found a function of shape of enclosure, type of nanoparticle suspended, particle size, Rayleigh number and the average temperature in the cavity. The effect of aspect ratio was also investigated by the authors and heat transfer enhancement was found higher for the 1.5–3.0. Akbari et al. [10] analyzed the single- and two-phase modeling of the laminar convection heat transfer in the $Al_2O_3$ nanofluids and found that the hydrodynamic fields predicted by both the modeling approaches were identical, while the thermal field predicted by the two-phase modeling were more accurate than the single-phase. However, the accuracy of thermal predication by the two-phase model decreased with the increased concentration and over-predicted the heat transfer coefficient of convection. Therefore, to recognize the causes responsible for the enhancement or deterioration in natural convection, a numerical simulation of natural convection at steady state in the distilled water and $Al_2O_3$—water nanofluids is conducted in the present study.

## 2 Physical Description of Problem

The test section is having the aspect ratio (width/height) of one, i.e., L = 120 mm, W = 120 mm and H = 120 mm. The physical boundary conditions for the test section are applied as shown in Fig. 1. The enclosure is heated from below by applying constant heat flux condition and cooled from above by the constant temperature boundary condition, with the adiabatic condition at walls, which makes the test section a Rayleigh-Bénard cavity. The objective of the present study is to find the augmentation of natural convection in a square cavity filled with nanoparticles suspension at the different particle concentrations.

## 3 Thermo-Physical Properties of Nanofluids

The working fluid in the present work are DW and $Al_2O_3$-DW nanofluid. The thermo-physical properties of DW have been taken from the Incropera et al. [11], while for nanofluids, proven analytical or empirical correlations are used, as shown below. The mixing theory was applied to estimate the density of the nanofluid in Eq. 1.

$$\rho_{nf} = (1 - \phi)\rho_{bf} + \phi\rho_p \tag{1}$$

Here, $\rho$—density, $\phi$—particle concentration, *p*—particle, *bf*, and *nf* are subscripts indicate base fluid and nanofluid, respectively. Khanaferet al. [12] formulated the correlations to estimate the expansion coefficient (Eq. 2) and specific capacity (Eq. 3) of the $Al_2O_3$-DW nanofluid.

**Fig. 1** Problem description of Rayleigh-Bénard cavity

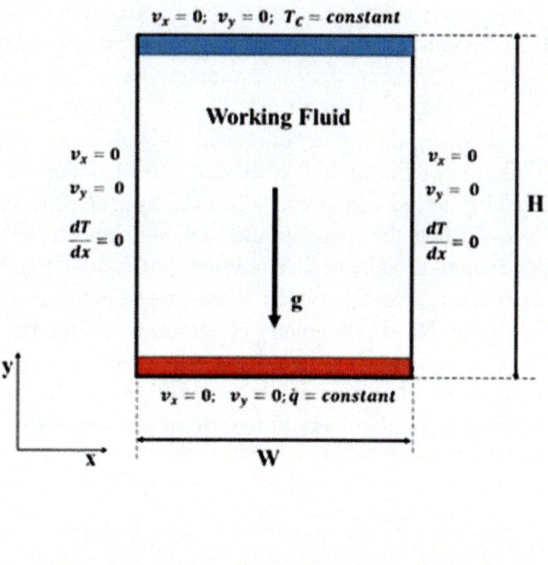

$$\beta_{nf} = \frac{(1-\phi)(\rho\beta)_{bf} + \phi(\rho\beta)_p}{\rho_{nf}} \tag{2}$$

$$c_{p,nf} = \frac{(1-\phi)(\rho c)_{bf} + \phi(\rho c)_p}{\rho_{nf}} \tag{3}$$

Here, $\beta$—thermal expansion coefficient and $c_p$—specific heat. Corcione [13] proposed the correlations to approximate the effective viscosity (Eq. 4) and effective thermal conductivity (Eq. 5), for the $Al_2O_3$-DW nanofluid.

$$\frac{\mu_{nf}}{\mu_{bf}} = \frac{1}{1 - 34.87\left(d_p/d_{bf}\right)^{-0.3}\phi^{1.03}} \tag{4}$$

Here $\mu$—dynamic viscosity, $d_{bf}$—equivalent diameter of the base fluid molecule.

$$\frac{k_{nf}}{k_{bf}} = 1 + 4.4\,\mathrm{Re}^{0.4}\mathrm{Pr}^{0.66}\left(\frac{T}{T_{bf,r}}\right)^{10}\left(\frac{k_p}{k_{bf}}\right)^{0.03}\phi^{0.66} \tag{5}$$

Here, $k$—thermal conductivity, Re—Reynolds number, $T$—temperature, and $r$—reference. The correlations given in Eqs. 4 and 5 are valid for the temperature in range of 20–60 °C and concentration in range of 0.002–0.09 vol. %. In the present study, the concentration of 0.01 and 0.1% is used, for which the above correlations are found suitable with a small loss of accuracy. Here, the dispersion of nanoparticles in the base fluid is only considered by thermo-physical properties.

## 4 Numerical Implementation and Grid Independency

The geometry shown in Fig. 1 is meshed in three different mesh strategies to divide the computational domain into the number of cells, as shown in Fig. 2.

For the present study, the pressure-based solver is found more suitable for the incompressible working fluid. The steady state with the atmospheric pressures and the gravitational acceleration with default values have been applied. As per the working conditions, the realizable model under the $k$-epsilon turbulence viscous model is found consistent with the physical concepts of the turbulent flow. As it assumed that the nanoparticles are uniformly suspended in the base fluid and a continuum is achieved. Therefore, instead of two-phase approach, single-phase approach is selected, which means, there is no traces of solid nanoparticles in the base fluid. No-slip boundary condition is assumed at all the walls of computational domain with constant heat flux at the heating surface (bottom wall) and constant temperature condition at cooling surface (top wall), respectively. For the spatial discretization, the body force weighted scheme is used for the pressure term,

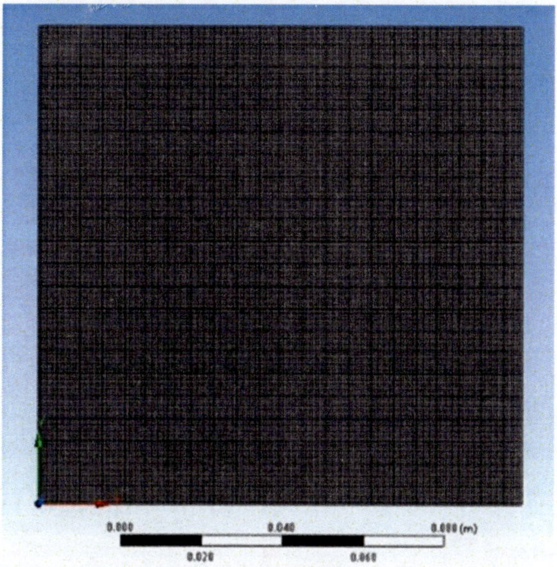

**Fig. 2** Mesh strategy for computation domain of problem

which is an ideal scheme for the buoyancy induced flows. The residuals of the continuity, velocity in x- and y-direction, turbulent kinetic energy is equal to the $10^{-3}$, while for the energy, the residuals for the energy is equal to the $10^{-6}$.

A grid independency test is conducted to find out the effect of mesh size on the results where if results are varying with the change in mesh, the results are not assumed to be correct. Three different mesh strategies are carefully chosen to find out the effect of mesh on the final output, i.e., *Nu,* as shown in Fig. 3. The calculated *Nu* from the numerical results of the simulation is almost same for all the mesh sizes with the maximum deviation of 4.80% from the average for mesh size 15655 × 15238, while the minimum deviation was obtained for the mesh size of 17644 × 17108. To obtain the accurate results with the minimum computing time and resources, and on the basis of independency test conducted, the mesh size of 17644 × 17108 is used for the simulation of heat transfer in DW and nanofluids.

## 5 Results and Discussion

In the present study, an experimental study conducted by the Zhou and Xia [14] to investigate the effect of heat flux on the Rayleigh-Bénard convection heat transfer is adopted for the validation purpose. The study reported the experimental results for the DW and formulated a correlation in the *Nu* and *Ra* for the above range of *Ra*, as given in Eq. 6.

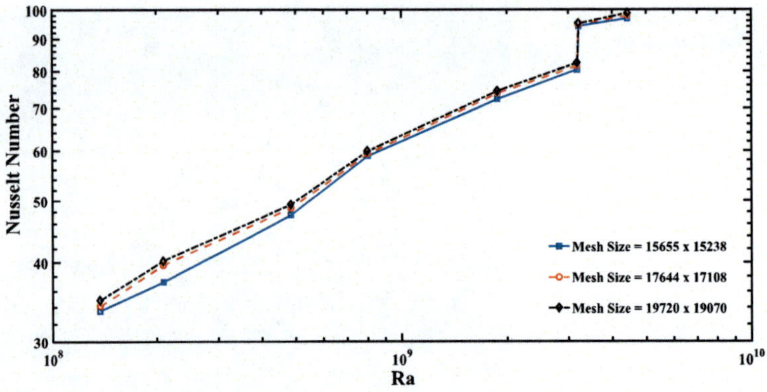


**Fig. 3** Grid independency check

$$Nu = 0.07 \pm 0.01\ Ra^{0.33 \pm 0.03}. \tag{6}$$

Another reason of selection of results reported by Zhou and Xia [14] was the range of *Ra*. The above correlation is valid for the range of *Ra* from $2 \times 10^7$ to $7 \times 10^{11}$, while the present numerical study is conducted for the range of *Ra* from $1.51 \times 10^8$ to $4.42 \times 10^9$. The comparison between the numerical results and experimental results of Zhou and Xia is plotted in Fig. 4. It can be observed from Fig. 4 that the numerical study under-predicted the *Nu* than that of experimental results, and the deviation is more for the higher range of *Ra*, beyond *Ra* $10^9$. By comparing both the results, a maximum deviation of 8.5% is occurred in the *Nu*, which is in the acceptable range of validation.

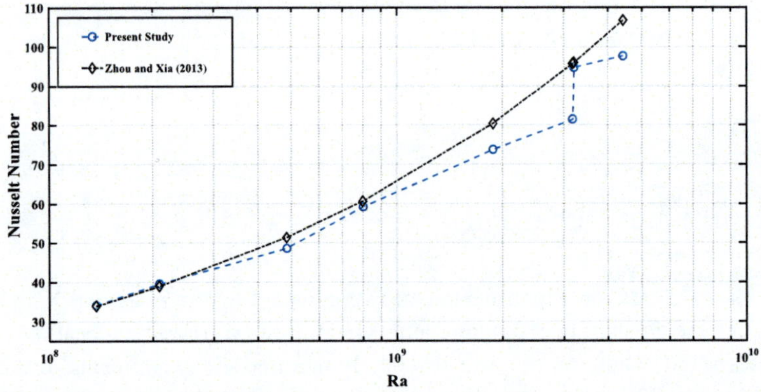


**Fig. 4** A semi-log plot between the *Nu* and *Ra* for DW calculated from the present study and Zhou and Xia [14]

The $Al_2O_3$ nanoparticles (diameter = 40 nm) are used in the present study with the concentration of 0.01 vol.% and 0.1 vol.%. In the experimental investigation conducted by Choudhary and Subudhi [15], a very small enrichment was experienced at the low *Ra* in the natural convection in nanofluid, and the opposite was found at the higher *Ra*. The similar results also occurred from the present numerical simulation, as shown in Fig. 5. At the $Ra < 6 \times 10^8$, the heat transfer enhancement occurred than that of the DW, while beyond this range of *Ra*, the nanoparticles suspension depreciates the heat transfer. The cause of this deterioration, reported by the Choudhary and Subudhi [15], was an unbalanced growth in the thermal conductivity as well as in the viscosity of the nanofluids. The primary advantage of the nanoparticle suspension is increased thermal conductivity, which further increased the ability of heat conduction. Contrary to the thermal conductivity increment, viscosity of nanofluid also increased which resists the fluid motion. The viscous diffusion is more due to nanoparticle presence in the base fluid, as compare to the thermal diffusion. As shown in Fig. 5, the present study is under predicting the heat transfer in the nanofluids as compared to the experimental results.

Isotherm and distribution of turbulent kinetic energy for the nanofluids of 0.01 vol.% particle concentration at is $Ra = 8.05 \times 10^8$ is shown in Fig. 6, respectively. Higher temperature is depicted at near the walls, whereas the convection central zone has lower temperature gradient. Thermal boundaries can also be observed at the cold top wall and hot bottom wall. Similar behavior is observed for the turbulent kinetic energy, which is also a function of temperature. As the temperature increases near the walls, higher velocity of the fluid particles occurred due to buoyancy force. The turbulent kinetic energy is function of the velocity of hot and cold plumes, occurred at the bottom and cold surface, respectively. Therefore, the turbulent kinetic energy increased at the hot sites, and about 304% variation in the turbulent kinetic energy occurred from at the walls to adjacent to the walls. At the walls, the turbulent kinetic energy is minimum due to no-slip condition. Another reason of the deterioration in heat transfer at the higher *Ra* can be observed from Fig. 7, where the velocity vectors are shown at $Ra = 8.05 \times 10^8$.

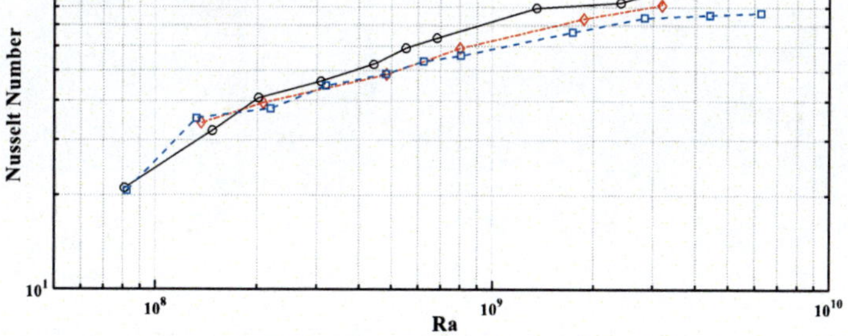


**Fig. 5** A log-log plot of the *Nu* with the *Ra* for the nanofluid (concentration = 0.01 vol.%)

At the left bottom and right top, two circulatory motions are visible, known as eddy currents, shown in the circles. These eddy currents create the resistance to flow of hot and cold plumes near to the solid surfaces. When the motion of hot and cold plumes is obstructed, heat transfer from area enclosed by the eddy currents is reduced, which leads to an overall reduction of heat transfer, as the eddy currents cover a large area of heat transfer. By increasing the *Ra* from $8.05 \times 10^8$ to $4.39 \times 10^9$, the velocity also increased from 0.017 to 0.031 m/s, as shown in Fig. 7. This velocity of fluid particles results in the two deviations, the location of the eddy currents changed and area covered by these eddy currents increased. The both deviations slow down the rate of transfer of heat from hot to the cold surface. At solid surface, the velocity of the fluid particles is zero, and at 1 cm from the horizontal solid surfaces, the velocity of fluid particles equal to 0.026 m/s.

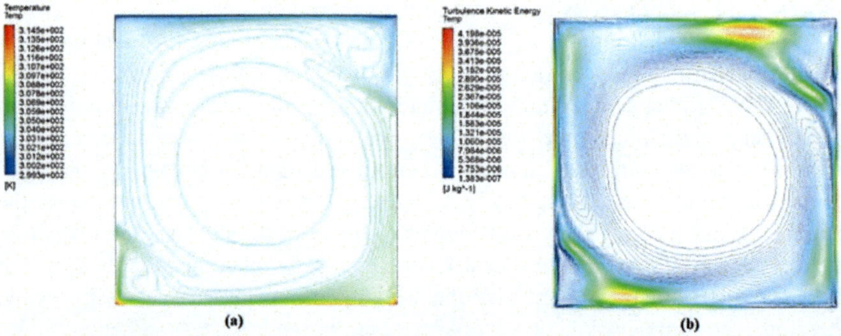


**Fig. 6** **a** Isotherms and **b** turbulent kinetic energy contours in the Rayleigh-Bénard convection at $Ra = 8.05 \times 10^8$ for nanofluids (particle concentration = 0.01 vol.%)

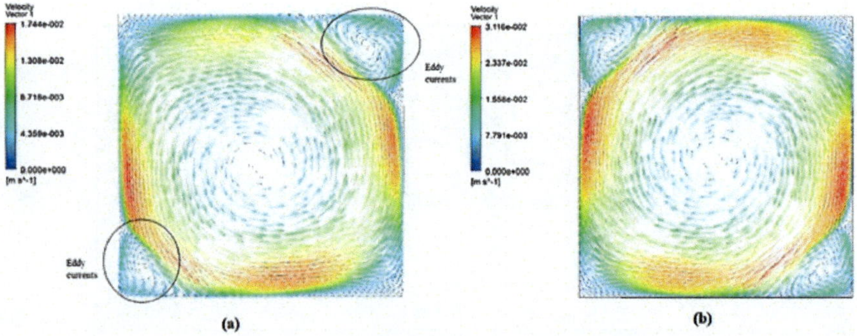


**Fig. 7** Velocity vectors in the Rayleigh-Bénard convection at **a** $Ra = 8.05 \times 10^8$ **b** $Ra = 4.39 \times 10^9$ for nanofluids (particle concentration = 0.01 vol.%)

The heat transfer is further enhanced by increasing the nanoparticle concentration up to 0.1 vol.%. At the low *Ra*, a high enhancement in the heat transfer occurred for higher particle concentration, as depicted in Fig. 8. While, the heat transfer again deteriorated after a critical *Ra*, $5 \times 10^8$, and the value of *Nu* is less than that of DW. For 0.1 vol.%, the results of numerical simulation are very near to the experimental results, a small deviation occurred. The eddy currents generated in a very large area are the reason of the deterioration in heat transfer, as shown in Fig. 9, which depicted the velocity vectors in the test section. In this case, a more resistance is generated to the plume motion. Choudhary and Subudhi [15] reported that the heat transfer deterioration due to increased viscosity with the particle concentration. At the same time, the Brownian motion of the nanoparticles increased with the temperature. This increased Brownian motion increased the rate of heat transfer, as a random movement of the nanoparticles carry the heat from hot

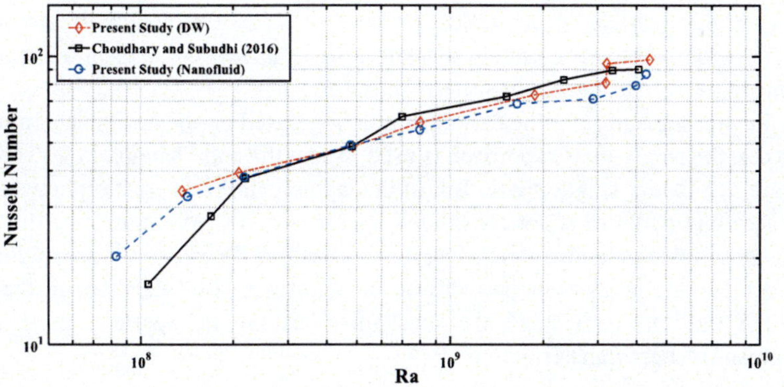


**Fig. 8** A log-log plot between the *Nu* and *Ra* for the nanofluids having particle concentration of 0.1 vol.%

**Fig. 9** Velocity vectors in the Rayleigh-Bénard convection at $Ra = 4.3 \times 10^9$ for nanofluids (particle concentration = 0.1 vol.%)

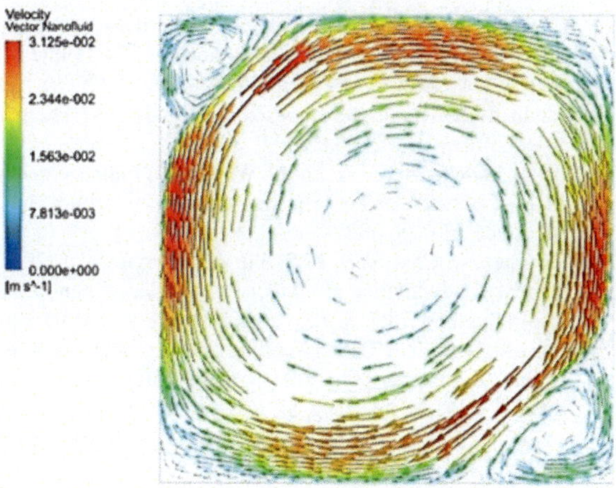

surface to the cold surface. But with the further increment in the Brownian motion agglomerates the nanoparticles and further increment in the viscosity occurred, which leads to degraded heat transfer than that of the DW.

## 6 Conclusions

Investigation on the heat transfer with the presence of nanoparticles in the Rayleigh-Bénard cavity is conducted in the present study. The size of nanoparticles used in the present study was 40 nm and two different particle concentrations were used to examine the use of $Al_2O_3$ nanoparticles. The results of the present simulation indicated the heat transfer improvement at $Ra < 8.5 \times 10^8$, while contrary trend occurred due to increased viscosity at higher *Ra* for both the concentrations of nanoparticles. The viscous diffusion was found dominated over the thermal diffusion for higher particle concentration. It was also found that eddy currents formed near the corners of the test section increased the resistance to the heat transfer, and the resistance further become strong with the *Ra*.

In the present study, for the ease of simulation based on the available computing resources, the single-phase approach is used, where the solid nanoparticles and base fluid are assumed a single phase. But in this approach, the interaction between the solid and fluid particles is not considered, that may be the reason of little deviation in the experimental and predicted results. It would be better to use two-phase approach for the analysis of suspension of nanosized solid particles in the conventional fluid to understand the mechanism of the nanoparticle motion and mechanism of heat transfer.

## References

1. Choi SUS, Eastman JA (1995) Enhancing thermal conductivity of fluids with nanoparticles. Int Mech Eng Congr Exhib, San Francisco, CA (United States)
2. Mintsa HA, Roy G, Nguyen CT, Doucet D (2009) New temperature dependent thermal conductivity data for water–based nanofluids. Int J Therm Sci 48:363–371. https://doi.org/10.1016/j.ijthermalsci.2008.03.009
3. Jana S, Salehi-Khojin A, Zhong WH (2007) Enhancement of fluid thermal conductivity by the addition of single and hybrid nano-additives. Thermochim Acta 462:45–55. https://doi.org/10.1016/j.tca.2007.06.009
4. Eastman JA, Choi SUS, Li S, Yu W, Thompson LJ (2011) Anomalously increased effective thermal conductivities of ethylene glycol based nanofluids containing copper nanoparticles. Appl Phys Lett 78:718–720. https://doi.org/10.1063/1.1341218
5. Das SK, Choi SUS, Patel HE (2006) Heat transfer in nanofluids—a review. Heat Transfer Eng 27:03–19. https://doi.org/10.1080/01457630600904593
6. Hwang KS, Lee JH, Jang SP (2007) Buoyancy-driven heat transfer of water-based $Al_2O_3$ nanofluids in a rectangular cavity. Int J Heat Mass Transf 50:4003–4010. https://doi.org/10.1016/j.ijheatmasstransfer.2007.01.037

7. Oztop HF, Abu-Nada E (2008) Numerical study of natural convection in partially heated rectangular enclosures filled with nanofluids. Int J Heat Fluid Flow 29(5):1326–1336. https://doi.org/10.1016/j.ijheatfluidflow.2008.04.009
8. Cianfrini C, Corcione M, Habib E, Quintino A (2014) Buoyancy-induced convection in $Al_2O_3$/water nanofluids from an enclosed heater. European J Mech B/Fluids 48:123–134. https://doi.org/10.1016/j.euromechflu.2014.04.014
9. Abouali O, Ahmadi G (2012) Computer simulations of natural convection of single phase nanofluids in simple enclosures: a critical review. Appl Therm Eng 36(1):1–13. https://doi.org/10.1016/j.applthermaleng.2011.11.065
10. Akbari M, Galanis N, Behzadmehr A (2011) Comparative analysis of single and two-phase models for CFD studies of nanofluid heat transfer. Int J Therm Sci 50(8):1343–1354. https://doi.org/10.1016/j.ijthermalsci.2011.03.008
11. Incropera FP, DeWitt DP, Bergman TL, Lavine AS (2006) Fundamental of heat transfer, 6th edn. Wiley Publication
12. Khanafer K, Vafai K (2011) A critical synthesis of thermophysical characteristics of nanofluids. Int J Heat Mass Transf 54:4410–4428. https://doi.org/10.1016/j.ijheatmasstransfer.2011.04.048
13. Corcione M (2011) Empirical correlating equations for predicting the effective thermal conductivity and dynamic viscosity of nanofluids. Energy Convers Manag 52:789–793. https://doi.org/10.1016/j.enconman.2010.06.072
14. Zhou Q, Xia KQ (2013) Thermal boundary layer structure in turbulent Rayleigh-Bénard convection in a rectangular cell. J Fluid Mech 721:199–224. https://doi.org/10.1017/jfm.2013.73
15. Choudhary R, Subudhi S (2016) Aspect ratio dependence of turbulent natural convection in $Al_2O_3$/water nanofluids. Appl Therm Eng 108:1095–1104. https://doi.org/10.1016/j.applthermaleng.2016.08.016

# Combustion and Performance Characteristics of Algae and Diesel Fuel Blends in a DICI Engine: An Experimental Approach

**Sumit Kanchan, Nihar Ranjan Swain, Rajesh Choudhary, and Patel CH**

**Abstract** In the present scenario of continuously increasing fuel cost, drop in import of the crude oil and inadequacy of raw crude oil in near future extends a great opportunity toward finding alternative oil. Among such alternatives, biodiesels promote maximum hegemony than other alternative energy sources. Third generation biofuels extend the advantages of being cultivated anywhere and consuming less field thereby giving high productivity. The primary contender in this way forward is algae. In this work, efforts have been made to analyze the combustion characteristics of *Schizochytrium sp.* algae oil with that of pure diesel. The fuel samples used in the study are pure algae biodiesel *(Schizochytrium sp.)* A100, A20 (20% algae biodiesel, 80% diesel) and pure diesel (ULSD- BSVI Oil). The test rig used single cylinder, 4-stroke, variable compression ratio (VCR), compression ignition, direct injection (DI) diesel engine. Combustion and performance parameters-like cylinder pressure, net heat release rate (NHRR), brake thermal efficiency (BTE), brake specific fuel consumption (BSFC) and volumetric efficiency were discussed. It was evident that average maximum cylinder pressure of algae biodiesel was 11.84% high than diesel. Further, both algae blends showed high-net heat release rate compared to diesel. It was also observed that algae biodiesel (A100) has higher volumetric efficiency (84.76%) at zero load, whereas at higher loads volumetric efficiency of diesel is higher.

**Keywords** Diesel · Algae biodiesel · Diesel engine · *Schizochytrium* specie · Combustion · Blend · Performance

## 1 Introduction

Setting years of fossils fuel and rise in cost envisioned researchers to explore renewable and sustainable fuels [1, 2]. Moreover, fossil fuels emit a group of carbon chain as well as nitrogen oxides, which proves carcinogenesis to human health [3].

S. Kanchan (✉) · N. R. Swain · R. Choudhary · P. CH
School of Mechanical Engineering, Lovely Professional University, Punjab 144411
Phagwara, India

 
R. Kumar et al. (eds.), *Recent Trends in Thermal Engineering*, Lecture Notes in Mechanical Engineering, https://doi.org/10.1007/978-981-16-3132-0_7

Algae are single cell organism found in fresh water conditions. One of the most important advantage of algae is that fuels can be derived from it for internal combustion applications. Algae productions are 15–300 times more than traditional crops because the growth rate is high within a short period. Moreover, $CO_2$ consumption during photosynthesis are also high than traditional crops due first rate of production [4–7]. This makes the algae a dearer for mass production. Saraswat et al. [8] investigated on a single cylinder 4-stroke VCR dual fuel research engine operated at constant 1500 rpm with fuel blends of butanol (B5, B10 and B15) and algae (A5, A10 and A15) and concluded that brake specific fuel consumption (BSFC) in case of butanol is greater than that of algae oil and gasoline due to lower calorific value. Brake thermal efficiency of butanol was found to be lesser than algae oil and that of gasoline for the reason that butanol has higher enthalpy of vaporization than gasoline. Kumar et al. [9] tested dual fuel blends of algae biodiesel, butanol and diesel in A5B25D70 and A10B30D60 on four strokes, four cylinder, Mitsubishi engine (4D34-2A) operated at 0–1700 Nm torque and 0–7500 rpm speed. It was found that BSFC of blends was better than that of pure diesel. Further, Sakthivel et al. [10] extracted raw biofuel from *Botryococcus braunii* algae through solvent extraction method and then by using transesterification process for converting raw biofuel to biodiesel. They added $SiO_2$ and $TiO_2$ additive for better performance. Diesel, B20, $B20TiO_2SiO_250$ and $B20TiO_2SiO_2100$ were used as testing fuel in a single cylinder CI at 1500 rpm and at various loads. It was concluded that BSFC and BTE were better in biodiesel blends. Further, Nirmala et al. [11] tested on 4-stroke Kirloskar TV1 vertical model IC engine operated at constant rpm and various load with waste cooking oil (WCO) and algal biodiesel with diesel. It was found that diesel performance was better than that WCO and algae blends. Deshmukh et al. [12] reviewed various type of oil extraction processes of algae and studied their performance and emissions.

In this current experiment, efforts have been made to analyze the combustion characteristics of *Schizochytrium sp.* algae oil with that of pure diesel. The test was conducted on a single-cylinder four strokes computerized test rig. The fuel samples used in the study are pure algae biodiesel (*Schizochytrium sp.*) A100, A20 (20% algae biodiesel, 80% diesel) and pure diesel (Ultra Low Sulfur Diesel-BSVI Oil).

## 2 Materials and Methodology

### *2.1 Biodiesel Production*

#### Fuel sample

Fuels used in the study are commercial diesel (compatible with BSIV vehicles) and algae oil of *Schizochytrium sp.* was purchased from local market of Delhi, India. The raw algae oil undergoes a transesterification process using biodiesel preparation

(a) (b) (c)

**Fig. 1** **a** Shows the setup used for the preparation of the biodiesel. The pure algae fuel formed is shown in (**b**) and the blend of algae with diesel is shown in (**c**)

unit comprising of reactor, condenser, chiller and hot water supply as shown in Fig. 1a. Fuel properties are measured using the standard devices for determining calorific value, viscosity, flash and fire point and acidity number. The values obtained is shown in Table 1. The calculated uncertainty in the measurement of above properties is found about 2% with the confidence level of 99% [13].

**Transesterification process**

Transesterification process is used for decreasing the free fatty acids of algae biodiesel. In this process, free fatty acids are removed from the raw algae oil, and the byproduct obtained is glycerol. Raw oil reacts with alcohol (methanol or ethanol) under the presence of KOH and $H_2SO_4$ to form biodiesel and glycerol. This process is cheaper due to abundant availability of catalyst [10]. Production biodiesel are shown in Fig. 2 In this experiment, as shown in Fig. 1 catalyst of NaOH (1% oil) mixed with ethanol ($C_2H_5OH$) (1:12 specific molar ratio with oil), after proper

**Table 1** Fuel properties

| Properties | Diesel | A100 | A20 |
|---|---|---|---|
| Calorific value (MJ/kg) | 45.734 | 40.9 | 44.767 |
| Density (kg/m3) | 785.24 | 843.65 | 796.92 |
| Kinematic viscosity @ 40 °C (cst) | 2.9 | 5.52 | 3.424 |
| Flash point ( °C) | 40 | 162 | 137.6 |
| Fire point ( °C) | 55 | 197 | 168.6 |
| Cloud point ( °C) | −6 | 8.5 | −3.1 |
| Pour point ( °C) | −12 | −8 | −11.2 |
| Acidity content (mg/ l $CaCO_3$) | 0.26 | 0.64 | 0.336 |

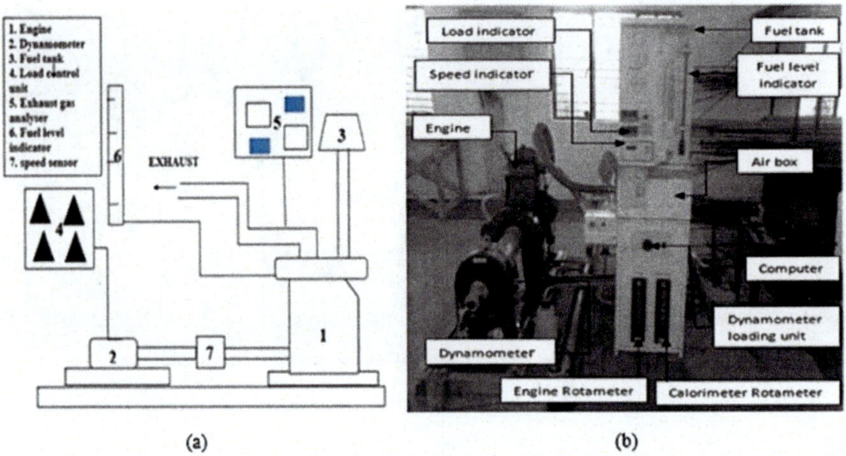


**Fig. 2** Schematic diagram of the experimental setup is depicted in (**a**) and the actual setup of the experimental test rig used is depicted in (**b**)

mixing, added with raw oil of algae. Reactor was pre cleaned with ethanol by stirring for five minutes. The mixture of raw oil is then dropped into reactor, maintaining it at 68 °C @ 600 rpm for up to 1 h. Reactor is settled for about 12 h. The esters are therefore separated from the glycerol using separator funnel. Ester is then washed with hot water @100 °C. Washed ester is than heated at 110 °C to make it water free and hence biodiesel was prepared.

### 2.2 *Preparation of Biodiesel Blend*

In this test A20 (80% diesel, 20% algae biodiesel) and pure algae biodiesel (A100) were used as blended fuel. A 20% of algae biodiesel and 80% of diesel (by v/v) were added in the glass bottle. The sample bottle was placed in a magnetic stirrer up to 1 h to make the mixture homogenous. Therefore, blend was prepared for testing. The fuel properties of the blend and pure algae are summarized in Table 1.

## 3 Engine Testing Setup

### 3.1 *Instrumentation and Engine*

In this experiment, single cylinder, 4-stroke, CI, water cooled, VCR, DI diesel engine are used. The details of the engine used are shown in Table 2. Figure 2a depicts the schematic and Fig. 2b shows the actual test rig used during the study.

**Table 2** Technical specifications of the engine used in the study

| | |
|---|---|
| Engine type | Variable compression ratio, 4-strokes, water cooled, single cylinder computerized diesel engine test setup |
| Make | Kirloskar |
| Type | 1 cylinder, 4-strokes, D.I diesel engine |
| Power | 3.5 KW @ 1500 rpm |
| Bore and stroke | 87.5 and 110 mm |
| Capacity | 661 CC |
| Compression ratio range | 12:1–18:1 (Tested at 18:1) |

Engine operates at constant speed (1500 rpm) with a constant compression ratio 18:1. The load was varied from 0 to 15 kg with an interval of 3 kg. Customized online software made by Apex India Manufacturers was used as the data acquisition system for the tests.

# 4 Results and Discussion

The initial trial with diesel fuel was made and thereafter with biodiesel and blends (A20 and A100). Performance and combustion parameters-like BTE, BSFC, volumetric efficiency, P-Θ diagram and net heat releasing rate were measured. The calculated uncertainty in the measurement of combustion and performance parameters is found about 3.4% with the confidence level of 99%.

## *4.1 Combustion Parameter*

**P-Θ diagram**

The variation in cylinder pressure of diesel, A20 and Algae biodiesel (A100) with crank angle is shown in Fig. 3. The increase in peak cylinder pressure is largely an indication of proper combustion which might be due to high amount of air mixing with fuel. From graph, it could be observed that maximum peak cylinder pressure rise has achieved for pure Algae fuel. Cylinder pressure of A100 is 11.84% higher than diesel. Cylinder pressure of diesel, A20 and algae biodiesel are 57.68 bar, 64.19 bar and 65.43 bar, respectively. This increase in cylinder pressure might be due to the fact that biodiesel fuels have high cetane number and has more density. Moreover, shorter ignition delay period could be another cause for this peak pressure rise.

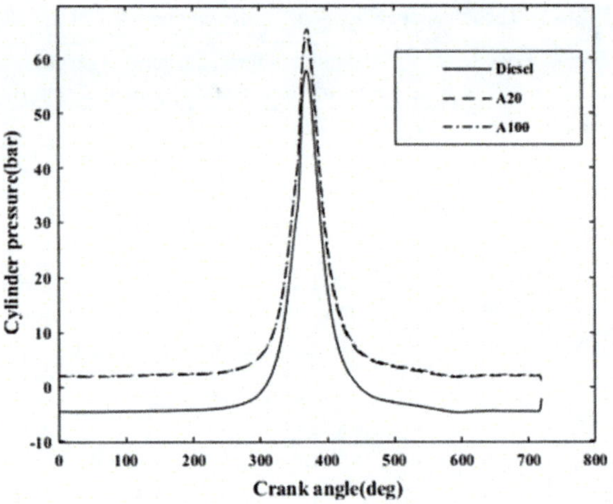


**Fig. 3** Result obtained for cylinder pressure variation with crank angle for diesel and biodiesel fuel samples

## Net Heat Release Rate

Figure 4 shows the net heat release rate variation with crank angle for diesel, A20 and algae biodiesel (A100) fuels. It can be observed that average maximum net heat release rate for diesel, A20 and algae biodiesel are 1 J/deg, 24.12 J/deg and 27.52 J/deg at 360° CA. The higher value of NHRR for A100 fuel might be due to the fact that biodiesels have higher oxygen content that may result in rapid combustion. Moreover, for A100 biodiesel fuel the NHRR value obtained is 12.35% higher than A20 blend. This difference is possible because of the fact that A100 contains more oxygen content than light blended (A20) fuel.

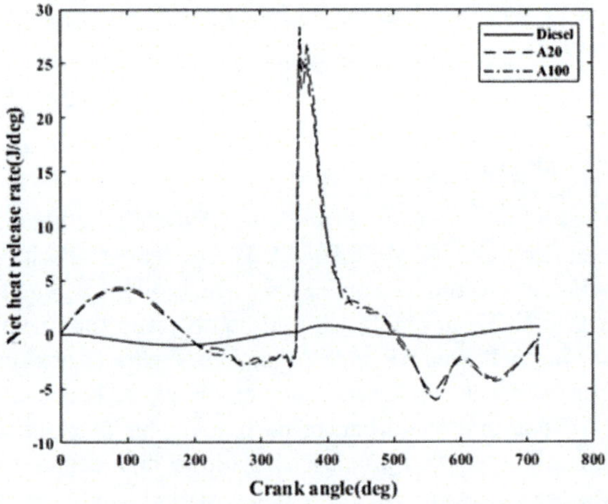


**Fig. 4** Result obtained for net heat release rate variation with crank angle for diesel and biodiesel fuel samples

## 4.2 *Performance Parameter*

**Brake Thermal efficiency**

Brake thermal efficiency (BTE) is a key performance parameter which is used for calculation of amount of useful energy available in the fuel energy application. Fluctuation in BTE at different engine loadings for algae biodiesel, A20 and Diesel is shown in Fig. 5. It can be observed that at full engine loading conditions, the values of BTE were found to be 29.14%, 30.58% and 31.75% for algae biodiesel, A20 and diesel, respectively. BTE of diesel fuel is maximum due to the fact that diesel has more calorific value and comparatively less viscosity than algae biodiesel blends.

**Brake Specific Fuel Consumption**

Amount of fuel required to produce 1 kWh power is called BSFC. Brake specific fuel consumption variation with load for diesel and biodiesel fuel samples is shown in Fig. 6. It is observed that BFSC of diesel and A100 algae blend at no load condition is 6.83 kg/kWh and 5.30 kg/kWh, respectively, while BSFC of A20 is 30.35 kg/kWh. This might be due to improper A/F mixing which give incomplete combustion initially. Then with increase in load BSFC gradually reduces up to full load. Diesel shows lower BSFC than algae biodiesel (A100) and A20. This might be possible due to less viscosity and higher calorific value of diesel than other biodiesel blends which are high-density biodiesel and consume extra fuel to obtain the same power output.

**Volumetric Efficiency**

Volumetric efficiency is the ratio between amount actual air consumed to the theoritical air consumption during suction stroke. Figure 7 shows the alteration in

**Fig. 5** Result obtained for brake thermal efficiency variation with load for diesel and biodiesel fuel samples

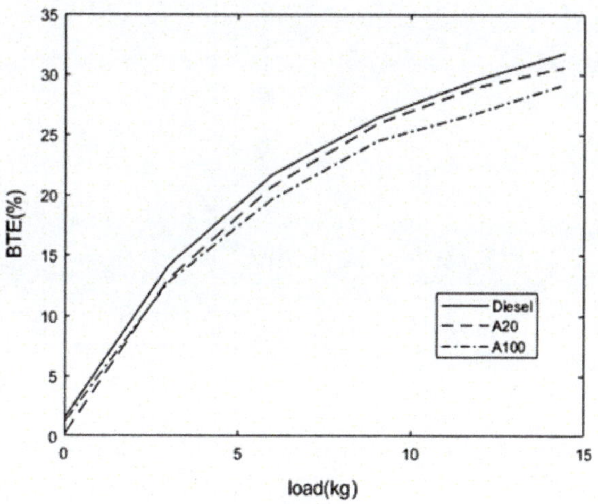

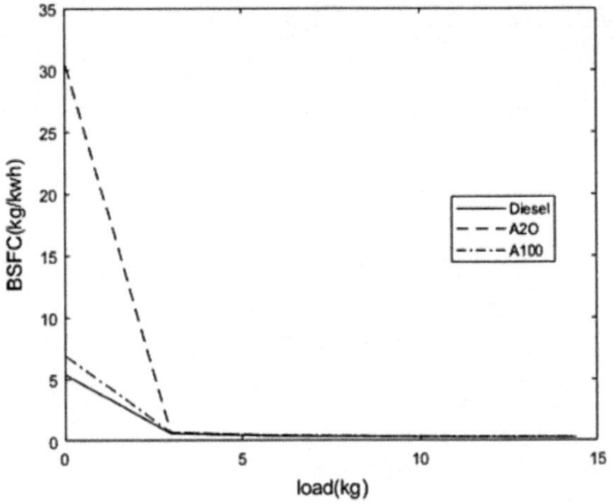


**Fig. 6** Result obtained for brake specific fuel consumption variation with load for diesel and biodiesel fuel samples

volumetric efficiency of A20, diesel and algae biodiesel at different engine load. Volumetric efficiency decreases with rise in engine load. It is evident from Fig. 7 that algae biodiesel (A100) has higher volumetric efficiency (84.76%) at zero load. This may be attributed to the fact that biodiesel blends are higher oxygen content fuels. Moreover, at higher loads the volumetric efficiency of diesel is more than that of A20 blend. This might be due to high density and more viscosity of biodiesel which creates improper air-fuel mixture inside the engine at higher loads.

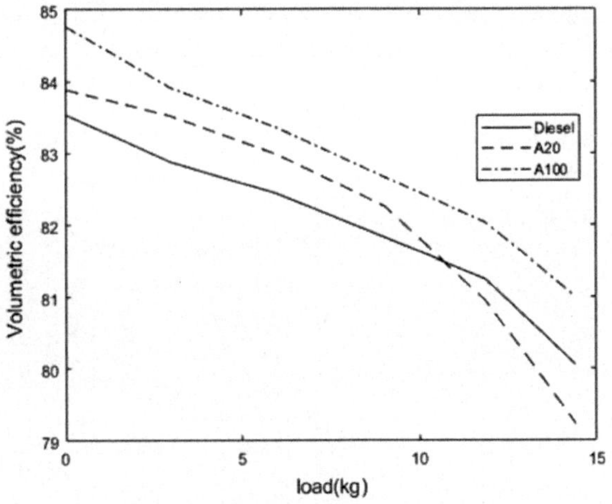


**Fig. 7** Result obtained for volumetric efficiency variation with load for diesel and biodiesel fuel samples

## 5 Conclusion

From the combustion and performance testing of A20, A100 and diesel fuel in 4-stroke single cylinder, CI, DI engine at varying loads and operating condition, it is evident that:

- Maximum peak cylinder pressure rise has achieved for pure Algae fuel. Cylinder pressure of A100 is 11.84% higher than diesel.
- For A100 biodiesel fuel, the NHRR value obtained is 12.35% higher than A20 blend.
- BTE of diesel fuel is maximum because diesel has higher calorific value and lighter viscosity than other algae biodiesel blends (A20 and A100).
- BFSC of A100 and diesel blend at no load condition is 6.83 kg/kWh and 5.30 kg/kWh, respectively, while BSFC of A20 is 30.35 kg/kWh.
- Algae biodiesel (A100) has higher volumetric efficiency (84.76%) at zero load whereas at higher loads volumetric efficiency of diesel is more than A20 blend.

## References


1. Piloto-Rodríguez R et al (2017) Assessment of diesel engine performance when fueled with biodiesel from algae and microalgae: an overview. Renew Sustain Energy Rev 69:833–842. https://doi.org/10.1016/j.rser.2016.11.015
2. Bauer N et al (2016) Assessing global fossil fuel availability in a scenario framework. Energy 111:580–592. https://doi.org/10.1016/j.energy.2016.05.088
3. Silverman DT et al (2012) The diesel exhaust in miners' study: a nested case–control study of lung cancer and diesel exhaust. J Nat Cancer Inst 104(11):855–868. https://doi.org/10.1093/jnci/djs034
4. Liu X et al (2017) Growth of Chlorella vulgaris and nutrient removal in the wastewater in response to intermittent carbon dioxide. Chemosphere 186:977–985. https://doi.org/10.1016/j.chemosphere.2017.07.160
5. Anto S et al (2020) Algae as green energy reserve: technological outlook on biofuel production. Chemosphere 242:125079. https://doi.org/10.1016/j.chemosphere.2019.125079
6. Schenk PM et al (2008) Second generation biofuels: high-efficiency microalgae for biodiesel production. Bioenergy Res 1(1):20–43. https://doi.org/10.1007/s12155-008-9008-8
7. Ullah K et al (2015) Assessing the potential of algal biomass opportunities for bioenergy industry: a review. Fuel 143:414–423. https://doi.org/10.1016/j.fuel.2014.10.064
8. Saraswat M, Nathi RC (2020) Comparative assessment of butanol and algae oil as alternate fuel for SI engines. Eng Sc Technol Int J 23(1):92–100. https://doi.org/10.1016/j.jestch.2019.04.002
9. Kumar V et al (2018) Production of biodiesel and bioethanol using algal biomass harvested from fresh water river. Renew Energy 116:606–612. https://doi.org/10.1016/j.renene.2017.10.016
10. Sakthivel R et al (2018) A review on the properties, performance and emission aspects of the third generation biodiesels. Renew Sustain Energy Rev 82:2970–2992. https://doi.org/10.1016/j.rser.2017.10.037

11. Nirmala N, Dawn SS, Harindra C (2020) Analysis of performance and emission characteristics of Waste cooking oil and Chlorella variabilis MK039712. 1 biodiesel blends in a single cylinder, four strokes diesel engine. Renew Energy 147:284–292. https://doi.org/10.1016/j.renene.2019.08.133
12. Deshmukh S, Ritunesh K, Kiran B (2019) Microalgae biodiesel: a review on oil extraction, fatty acid composition, properties and effect on engine performance and emissions. Fuel Process Technol 191:232–247. https://doi.org/10.1016/j.fuproc.2019.03.013
13. Singh S, Anuj J, Sunil Kumar M (2020) Sampled-data model validation: an algorithm and experimental setup of dual fuel IC engine. Fuel 279:118517. https://doi.org/10.1016/j.fuel.2020.118517

# Thermal Performance Assessment of Greenhouse Solar Dryer Operated Under Active Mode

**Vipin Shrivastava, Anil Singh Yadav, and Nitin Shrivastava**

**Abstract** In the present paper, an even span roof-shaped greenhouse dryer's thermal performance has been experimentally evaluated under no-load condition. The objective of the work is to evaluate the various heat transfer parameters inside greenhouse dryer so that proper crops would be selected for drying purposes. The average set value of air mass rate used for evaluation is 0.14, 0.21 and 0.28 kg/s for each consecutive days of operations. The results showed that the optimum value of the mass flow rate of air at which heat loss is less and heat gain is more.

**Keywords** Greenhouse solar dryer · Heat loss · No load · Active mode · Heat gain

## Notation

| | |
|---|---|
| $A_{fl}$ | Floor area ($m^2$) |
| $B_{dr}$ | Breadth of the dryer (m) |
| $C_p$ | Specific heat at constant pressure (J/kgK) |
| $h_{ca}$ | Convective-heat-transfer coefficient of the canopy-ambient ($W/m^2K$) |
| $h_{ra}$ | Radiative-convective-heat-transfer coefficient ($W/m^2K$) |
| $I$ | Sun intensity falling on greenhouse ($W/m^2$) |
| $K$ | Thermal conductivity (W/mK) |
| $P_T$ | Partial vapor pressure at temperature $T$ ($N/m^2$) |
| $T$ | Time in seconds |
| $X$ | Characteristic length (m) |
| $T_{grd}$ | Temperature of ground (°C) |
| $T_r$ | Temperature of room (°C) |
| $T_o$ | Ambient temperature (°C) |
| $U$ | Overall heat transfer coefficient ($W/m^2K$) |

V. Shrivastava (✉) · A. S. Yadav
Lakshmi Narain College of Technology, Bhopal 462022, India

N. Shrivastava
University Institute of Technology, Rajiv Gandhi Proudyogiki Vishwavidyalaya, Bhopal 462033, India



R. Kumar et al. (eds.), *Recent Trends in Thermal Engineering*, Lecture Notes in Mechanical Engineering, https://doi.org/10.1007/978-981-16-3132-0_8

$V_a$ Velocity of air inside the dryer (m/s)
$V_f$ Velocity of fan (rpm)

**Subscripts**

$\Phi$ Relative humidity of air (%)
$\sigma$ Stefan-Boltzmann constant ($W/m^2k^4$)
$\epsilon$ Emissivity
$c$ Cover

## 1 Introduction

Food loss is a major concern of the world and is increasing day by day. It has been estimated that 486 and 392 million tons of vegetables and fruits are produced annually throughout the global, respectively [1]. Solar drying is ancient and probably the first-ever food preservation method [2]. Drying can be done through open sun drying, industrial drying or solar drying. Sun-drying has its limitation. Industrial processes use conventional fuels as their heat sources. The third option is the most suitable one, which is solar drying. Among different categories of the dryer, a greenhouse is the best dryer in which vegetable and cash crops can be easily dried [3]. In this paper, greenhouse solar dryer is operated under forced convection mode. The readily available plastic cover used on the floor to increase the temperature inside the dryer. The dryer is tested under no-load condition so that a suitable crop would be selected for drying. For this purpose, it is necessary to the determination of various thermal parameters. The research on the coefficient of performance (COP) and heat utilization factor (HUF) for greenhouse dryers (GD) in unloading conditions is rarely seen.

## 2 Experimental Setup

A low cost even span solar greenhouse dryer is erected at the Energy Park of institute (23.25° N–77.5° E). The dryer has been oriented in the direction of the east-west during the experiment for extracting maximum solar radiation. The schematic diagram of greenhouse solar dryer is shown in Fig. 1. The sidewalls of the setup are inclined at 23.2°. The effective floor area of the dryer is 4.54 $m^2$ and the central height being 3 m. The greenhouse dryer is covered with a 6-mm thick polycarbonate sheet. Openings that are square in shape and having dimensions 0.24 m × 0.24 m each have been provided at the top of the dryer's backside wall for air circulation inside the dryer. An AC fan fit at the top of the back wall.

**Fig. 1** Photograph of greenhouse solar dryer setup

The experimental values for ground temperature were first determined under the uncovered floor condition for three days from 11th to 13th September 2020. The mass flow rate of air was kept to be 0.14 kg/s for the first day, 0.21 kg/s for the second and 0.28 kg/s for the final day. A plastic cover of thickness 0.4 mm was used for keeping the floor covered during the experiment.

## 3 Instrumentation

A calibrated J type thermocouples measure the temperature at various points. Humidity inside the greenhouse solar dryer is measured by a calibrated probe type hygrometer. The range and accuracy of the hygrometer are 10–95% and 0.1%. Incident insolation is assessed by the calibrated solar kit, which has the range of 0–1999 W/m$^2$. A calibrated digital-anemometer is measure the velocity of air. The range and accuracy of anemometer are 0.4–30.0 m/s and ±2%, respectively.

## 4 Numerical Computation

The value of ground-room coefficient signifies the amount of heat flux occurring here per unit area per unit temperature difference. Convective-heat-transfer coefficient from the ground to a room is calculated by following relation [4]

$$h_{gdr} = 0.884\left(T_{gd} - T_{rm} + \frac{\left[P(T_{gd}) - \phi P(T_a)\right](T_{gd} - 273)}{268.9 \times 10^3 - P(T_{gd})}\right)^{1/3} \tag{1}$$

where $T_{gd}$ and $T_{rm}$ are ground and room temperature (°C).

The radiative heat transfer coefficient from room to ambient is calculated by the following relation [5].

$$h_{ra} = \frac{\sigma\varepsilon\left[(T_{gd} + 273.15)^4 - (T_{rm} + 273.15)^4\right]}{(T_{gd} - T_{rm})} \tag{2}$$

Garzoli and Blackwell [6] proposed the relation for convectively transfer of heat from canopy in case of greenhouse solar dryer ($h_{ca}$)

$$h_{ca} = 7.2 + 3.8V_w \tag{3}$$

The overall heat transfer coefficient ($U$) is determined as

$$\frac{1}{U} = \frac{1}{h'} + \frac{X}{K_c} + \frac{1}{h_{ca}} \tag{4}$$

Value of $h'$ and $h_{ca}$ is calculated with the help of Eqs. (5) and (3) as:

$$h' = h_{gdr} + h_{ra} + h_{ev} \tag{5}$$

The value of $h_{ev}$ has been neglected because for the forced convection in an unload conduction, the evaporative losses are very less.

HUF is a measure of a drop in temperature due to cooling air during drying and an increase in temperature due to air temperature [7]. COP is the ratio of temperature difference of room-ambient and ground-ambient. These two parameters have opposite nature to each other as one is increasing and the second decreasing and thus crossing each other at a value [8].

$$\text{HUF} = \frac{T_{rm} - T_{gd}}{T_{rm} - T_o} \tag{6}$$

and

$$\mathrm{COP} = \frac{T_{\mathrm{gd}} - T_{\mathrm{o}}}{T_{\mathrm{rm}} - T_{\mathrm{o}}} \tag{7}$$

## 5 Result and Discussion

Figure 2 depicts the changes in environmental parameter of consecutive days. The weather conditions were nearly the same for each experimental day, and the sky is cloud-free. Environmental humidity is in the range of 23–29%.

The disparity of ground-surface temperature in active flow is depicted in Fig. 3. The maximum ground temperature is observed at 41 °C at the lowest air mass flow rate. Ground temperature is an important parameter to determine convective-heat-transfer ($h_{ra}$) from room to air. The average value of $h_{gdr}$ and $h_{ra}$ were calculated to be 4.69, 4.66, 4.55 W/m$^2$K and 6.24, 6.17, 6.04 W/m$^2$K for day 1, 2 and 3, respectively.

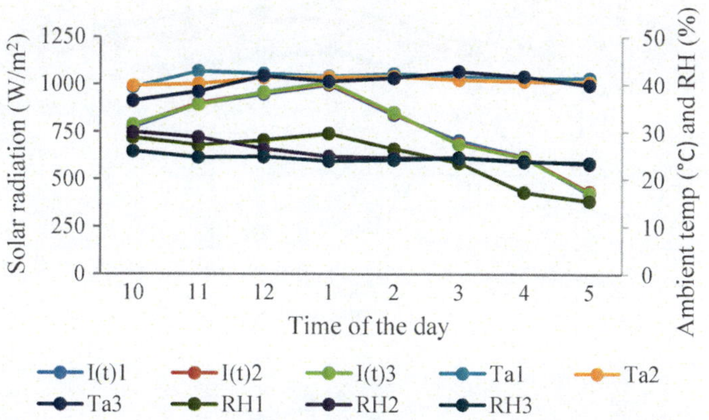


**Fig. 2** Variation of solar radiation, ambient temperature and RH on different days of experiment

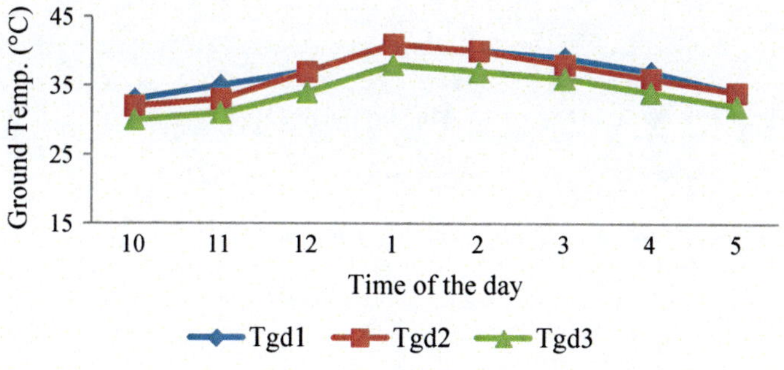


**Fig. 3** Variation of ground temperature on different days of the experiment

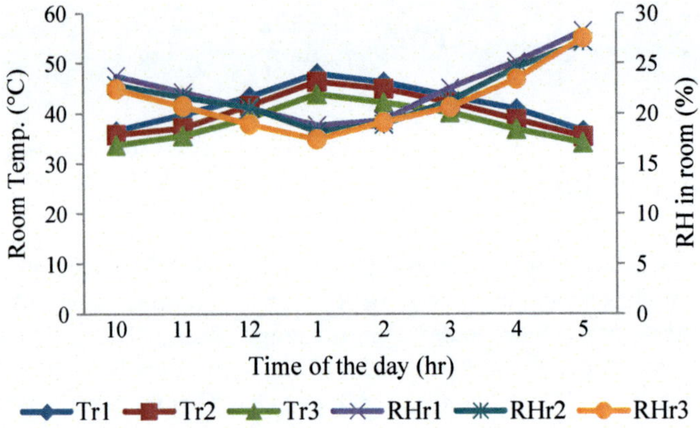


**Fig. 4** Changesin temperature and RH of GD

(a)

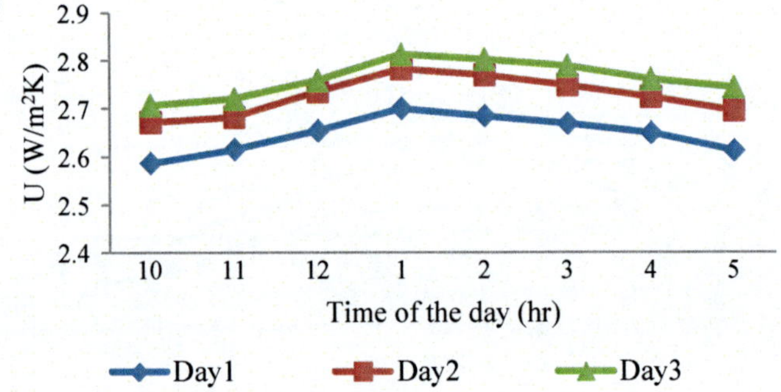


(b)

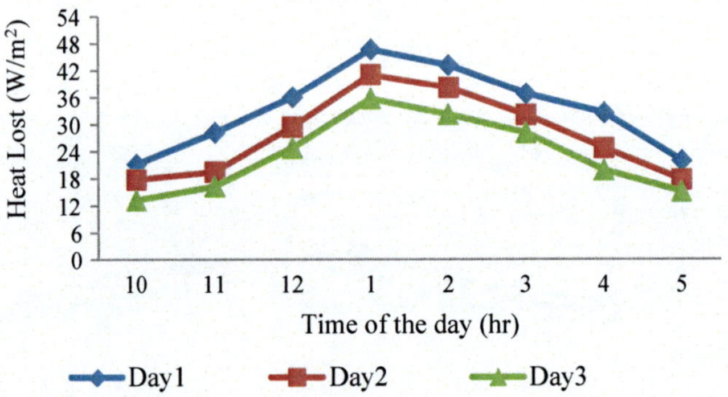


**Fig. 5** **a** Variation of overall heat loss coefficient and **b** heat loss

The changes in room's temperature and percentage humidity is shown below in Fig. 4. The maximum room temperature in covered condition on the consecutive experimental days is recorded to be 47.9 °C, 46.3 °C and 43.8 °C, respectively. Under unload condition temperature and humidity are inversely proportional. Data shows that a constructed dryer is best suitable for paddy, wheat, corn, rice drying.

The variation in resulted heat transfer parameters is presented in Fig. 5a, b. The average value of overall heat transfer coefficient on sequential days were calculated

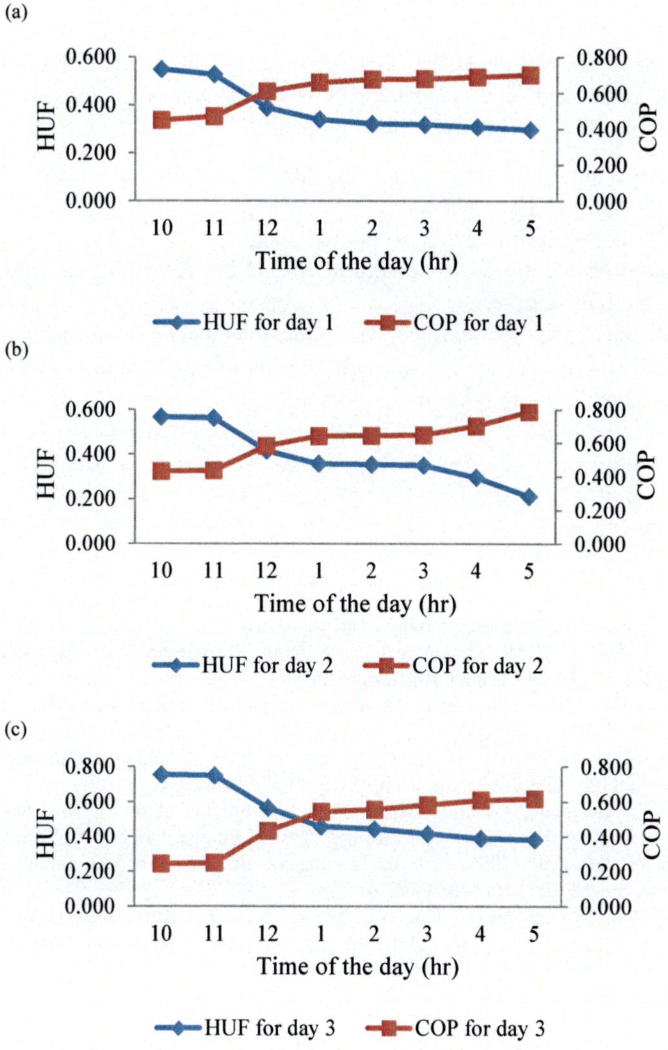


**Fig. 6** Variation of HUF and COP for **a** first, **b** second and **c** third day

to be 2.65, 2.73 and 2.78 W/m$^2$K, respectively. This overall heat loss coefficient at different mass flow rates further used to make thermal modeling of greenhouse solar dryer in other conditions. But heat loss is proportional to air movement. Since heat loss depends on the value of room temperature and the room temperature from day 1 to day 3 is getting decreased, so heat loss is also found to be decreased.

The average heat utilization factor is found maximum in third day. It is inversely proportional to the heat losses and is shown in Fig. 6.

## 6 Conclusion

A general and simple model of solar greenhouse dryer were developed. The experiment performed in varying mass flow rate condition in forced convection mode for three days reveals the following results:

- The temperature in the dryer at the peak hour is higher than outside temperature by 17.3, 14.8 and 12.7 °C.
- Heat loss is decreasing and HUF is increasing.
- The data obtained at unload condition will helpful for drying of crop and predicting the temperature.
- The calculated values of various thermal parameters can be utilized in selecting a suitable crop for drying applications. Greenhouse dryer is the best suitable for paddy, wheat, corn, rice drying.

## References

1. Chauhan PS, Kumar A (2017) Heat transfer analysis of north wall insulated greenhouse dryer under natural convection mode. Energy 118:1264–1274
2. Gupta PM, Das A (2017) Design and construction of solar dryer for drying agricultural products. Int Res J Eng Technol 4(3):1946–1951
3. Almuhanna EA (2012) Utilization of a solar greenhouse as a solar dryer for drying dates under the climatic conditions of the eastern province of Saudi Arabia. J Agric Sci 4(3):237–246
4. Sukhatme SP, Nayak JK (1993) Solarenergy: principles of thermal collection and storage, 3 edn. Tata McGraw Hill Publishing Company Limited, New Delhi
5. Mishra SP, Shrivastava V (2017) Experimental investigation of one glass cover and three absorber plates two-pass solar air collector with thermal storage. Indian J Sci Technol 10:1–10
6. Garzoli KV, Blackwell J (1981) An analysis of the nocturnal heat loss from a single skin plastic greenhouse. J Agric Eng Res 26:203–214
7. Tiwari GN (2009) Fundamental of solar dryers. Anamaya Publishers, New Delhi
8. Brenidorfer B, Kennedy L, OswinBateman CO, Trim DS, Mrema GC, Wereko-Brobby C (1986) Solar dryers: their role in post harvest processing. Commonwealth Secretariat Marlborough House, London

# Effect of Copper Fins on Fresh Water Productivity of Pyramid Solar Still

**Naveen Sharma, Shaik Noushad, and G. Siva Ram Kumar Reddy**

**Abstract** This research work presents an experimental performance assessment of a pyramid solar still (PSS). The square type of PSS is developed and tested under the environmental conditions of Kanchikacherla (16.6834°N, 80.3904°E), Andhra Pradesh, India. It is used to produce the fresh water and hot water simultaneously from the brackish feed water for domestic applications. Conceptually, it is expected that the rate of evaporation is higher when the surface area for heat absorption and rejection is more. Therefore, the pyramid solar still has been tested for smooth absorber plate (CASE-1) and the absorber plate with copper fins positioned horizontally at 140 mm (CASE-2) and 90 mm (CASE-3) apart. Experiments have been carried out between 7:00 AM and 7:00 PM, only on sunny days, for reducing the impact of variation in solar radiation intensity. It has been observed that the amount of distillate yield strongly depends upon the solar radiation intensity and ambient temperature. The introduction of copper tubes, which act as fins, on the black absorber plate leads to increase in daily fresh water production, up to a maximum value of 1.95 $L/m^2$ for CASE-3. The distilled water output for CASE-3 and CASE-2 is about 60% and 33% higher than that of CASE-1, respectively. Additionally, the water quality (WQ) parameters such as total dissolved salts, pH, and electrical conductivity ($dSm^{-1}$) of the produced fresh water are measured as 28 mg/L, 7.02, and 0.08 $dSm^{-1}$, respectively, which meet the WQ-standards as suggested by the World Health Organization.

**Keywords** Solar energy · Solar still · Water distillation · Fins · Distillate output

N. Sharma (✉)
Department of Mechanical Engineering, Netaji Subhas University of Technology, Dwarka, Delhi 110078, India

N. Sharma · S. Noushad · G. Siva Ram Kumar Reddy
Department of Mechanical Engineering, DVR & Dr. HS MIC College of Technology, Kanchikacherla, Andhra Pradesh 521180, India

 
R. Kumar et al. (eds.), *Recent Trends in Thermal Engineering*, Lecture Notes in Mechanical Engineering, https://doi.org/10.1007/978-981-16-3132-0_9

# 1 Introduction

Renewable energy sources have the potential to overcome many burning problems of mankind such as global climate change, environmental degradation, drinking water, and energy scarcity, etc., and their efficient use can lead to sustainable development. Solar energy, one of the cleanest, cheapest, and most abundant renewable energy source, is effectively harnessed for a variety of applications for instance power generation, solar desalination/distillation, solar water/air heating, solar drying, solar cookers and heating/cooling of buildings [1–4].

Worldwide, most of the countries are suffering from fresh water scarcity owing to industrialization and increasing population. Specifically in South India, the shortage of fresh water poses a big threat to the humanity, where most of the available water (underground or sea water) is not appropriate for household needs, i.e., drinking and cooking. Additionally, the contamination of potable assets, such as rivers, wells, lakes, and underground water, by industrial wastes has further raised the concern. Among various ways of water purification, solar distillation using solar stills appears as one of the promising technology to produce bulk amount of fresh water from saline water or sea water [5, 6]. These devices are suitable for serving water demand for small communities in remote areas that are not having conventional facility. Solar still uses evaporation and condensation heat transfer mechanisms for producing the fresh water from brackish water using solar energy. But, the fresh water productivity of conventional solar still is usually low; therefore, various investigations have been carried out by changing the design of solar still or hybrid solar stills with flat plate/parabolic collector, or using fins and heat storage materials (latent/sensible), etc. to improve the yield of solar still [7–16].

Fathy et al. [7] reported that the double slope solar still coupled with PTC tracking system produces higher freshwater about 28.1% than solar still with fixed PTC and about 142.3% than conventional solar still (CSS), at a water depth of 20 mm during the summer conditions. The findings of Yousef and Hassan [8] revealed that the productivity of single slope solar still (SSSS) increases by 9.5% using PCM, 17% using pin fins and PCM, 13% using steel wool fiber and PCM, and 25% using steel fibers compared to CSS. Sakthivel and Arjunan [9] tested the performance improvement of a SSSS using cotton cloth energy storage medium under the climatological conditions of Chennai, India. The outcomes of this study revealed that the fresh water productivity is augmented by 24.1% than CSS for 6 mm cotton cloth thickness. Arunkumar et al. [10] investigated the influence of CuO nanostructured coated absorber plates (NCAP) combined with polyvinyl alcohol (PVA) sponges on the fresh water productivity of a SSSS. The improvement in yield of fresh water is about 40% with SSSS–CuO–NCAP and 8% with SSSS–CuO–NCAP–PVA sponges, while about 8% reduction in productivity found for SSSS–PVA sponges. Panchal et al. [11] observed that the fresh water productivity is improved by 113.52% and 104.68%, respectively, for solar still integrating evacuated tubes and solar still incorporating both evacuated tubes and calcium stones than that of CSS. Bataineh and Abbas [12] found that the efficiency of solar still is enhanced up to 46% with addition of internal reflectors in South, East and West sides in December.

The productivity of double-basin solar stills with two different wick materials, namely jute and cotton cloth, has been evaluated by Modi and Modi [13]. Maximum amount of yield is obtained at 0.02 m level of water depth and enhancement is found to be 21.46% in case of jute cloth than that of black cotton cloth. Effect of extended surfaces, i.e., circular and square hollow fins, on the productivity of a double slope single basin solar still is explored by Jani and Modi [14]. Water depth is varying from 10 to 30 mm at an interval of 10 mm. The maximum distilled output of 1.4917 $kg/m^2$-day is obtained for solar still with circular fins, while the maximum yield of 0.9672 $kg/m^2$-day is observed for square fins still at a typical water depth of 10 mm.

Kabeel et al. [15] examined the performance improvement of PSS by embedding array of hollow circular fins over the absorber plate along with a PCM tank just underneath the absorber surface. The daily productivity is improved by 43% and 101.5%, respectively, for solar still with hollow fins and solar still with PCM addition compared to CSS. The influence of fins arranged in vertical and inclined manner on the productivity of a SSSS has been studied by Panchal et al. [16] under the environmental conditions of Gandhinagar, Gujarat. Experimental results revealed that the improvement in daily productivity of 26.77% and 24.19% is observed for inclined and vertical fins, respectively, than that of CSS.

From literature review, it is clear that the studies on PSS with copper fins mounted horizontally on the absorber plate have not been addressed adequately. In this regard, the present work makes an attempt to investigate the effect of placing cooper tubes as fins and their spacing on the productivity of square PSS.

## 2 Experimental Facility and Instrumentation

The schematic illustration of the pyramid type solar still is depicted in Fig. 1. The experimental setup has been fabricated at the Department of Mechanical Engineering, DVR & Dr. HS MIC College of Technology, Kanchikacherla, India.

The still basin is made of galvanized iron sheet with inner dimensions of 0.710 m × 0.710 m and depth of 22.5 cm at the center. The basin/absorber plate sheet is painted black for absorbing the maximum solar radiation. The condensing glass cover is fabricated in the pyramid shape in order to increase the surface area for vapor to come in contact. Plain glass, having thickness of 3.5 mm and transmissivity 0.9, is used as glass cover, and the glass tilt angle is fixed at 16.68° to the horizontal surface. The side walls and the base have been made of wood having thickness of 22 mm.

To improve the performance of PSS, copper tubes fins (diameter = 20 mm, length = 710 mm) are mounted horizontally at the spacing of 140 mm (CASE-2) and 90 mm (CASE-3) over the absorber (see Fig. 2). The spacing between copper

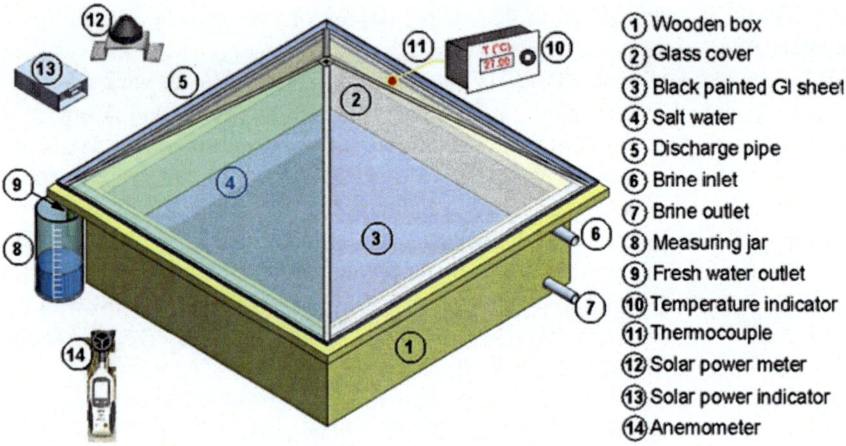


**Fig. 1** 3D model of fabricated pyramid type solar still with instrumentation

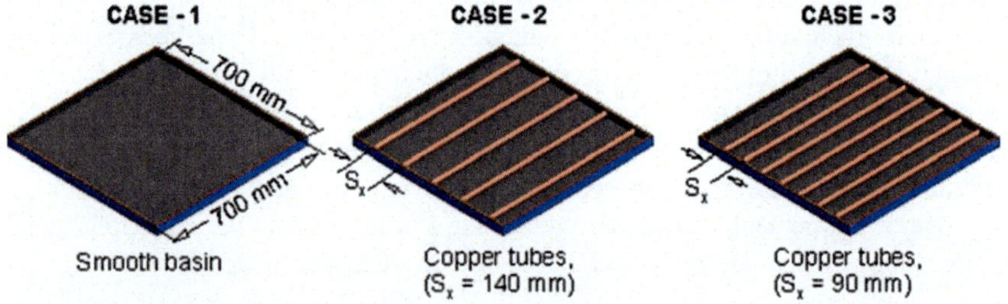


**Fig. 2** Configurations considered for improving yield of pyramid solar still

tubes has been selected judiciously so as to get noticeable improvement in fresh water yield and the copper tubes may not cause shading effect on absorber surface.

Firstly, the basin is filled by the saline water to a particular level (water depth of 3 cm) from the bottom. The incident energy is transmitted to the water through the glass and absorbed by the black absorber plate. Once water starts heating and absorbs enough amount of energy, then water vapor escapes at the interface of liquid surface. The generated vapor travel to the condenser section and gets condensed due to temperature difference at the glass surface. The condensate mass slides over the inclined glass and can be collected in a measuring jar. The purification takes place inside the solar still through evaporation and condensation process; therefore, the condensate obtained is suitable for household needs.

In the present work, calibrated K-type thermocouples in conjunction with a digital temperature indicator are used for measuring the temperature data such as basin temperature, wall temperature, water temperature, vapor temperature, glass inner and outer temperatures at a time interval of 30 min. Further, an anemometer and a solar power meter are employed to measure the wind velocity and solar

**Table 1** Accuracy, range, and error of measuring instruments

| Instruments | Range | Accuracy | Standard error |
|---|---|---|---|
| Solar power meter | 0–1800 W/m$^2$ | ±5 W/m$^2$ | ±2.88 |
| Air flow anemometer | 1–25 m/s | ±0.2 m/s | ±0.12 |
| Thermocouple | 0–275 °C | ±1 °C | ±0.58 |
| Digital temperature indicator | 0–1000 °C | ±1 °C | ±0.58 |
| Measuring jar | 0–1000 ml | ±50 ml | ±28.87 |
| pH meter | 0–14 pH | ±0.01 | ±0.006 |
| Total dissolved salts (TDS) | 0–999 PPM | ±2 PPM | ±1.15 |
| Turbidity meter | 0–1000 NTU | ±1 NTU | ±0.58 |
| Conductivity meter | 200–1000 μs/cm | ±0.1 μs/cm | ±0.058 |

radiation intensity, respectively. In addition, the quality of saline and purified water is measured by a pH meter, a TDS meter and a conductivity meter. The range, accuracy, and standard uncertainty of the measuring instruments are illustrated in Table 1.

## 3 Results and Discussion

The experiments are carried out from 7.00 AM to 7.00 PM for measuring the temperatures after every half an hour, at suitable locations, and hourly production of fresh water. The variation of solar radiation intensity on the days of experimentation (February 18–23, 2020) for all cases is presented in Fig. 3. As expected, solar radiation intensity gradually increases from 7:00 AM, attains the maximum value of 936 W/m$^2$ at the noontime, i.e., 12:00 PM, and thereafter decreases with further increase in time and becomes zero after 6:00 PM due to absence of sunlight. It is confirmed that the highest variation in solar radiation intensity is observed to be within ±5%. Therefore, the experimental data of fresh water yield obtained for different cases can suitably be compared in order to find the effectiveness of introducing copper tubes inside PSS.

The still temperatures, mainly ambient temperature, water temperature, inner glass temperature, and basin temperature, are of prime importance in order to understand the results of fresh water productivity and shown in Fig. 4 for all the cases. At the beginning of the day, all the temperatures are increasing first and reach the maximum, around 1:30 PM, and afterward it will decrease. Though, a time lag between the maximum values of temperature and solar radiation intensity is observed that can be associated with the time taken in heating the solar still parts.

For all the cases, water temperature has the highest value followed by basin temperature, glass temperature and ambient temperature. The maximum temperatures of salty water inside the PSS for CASE-3, CASE-2 and CASE-1 are about 63 °C, 59 °C, and 56.5 °C, respectively. The copper tubes have a high thermal

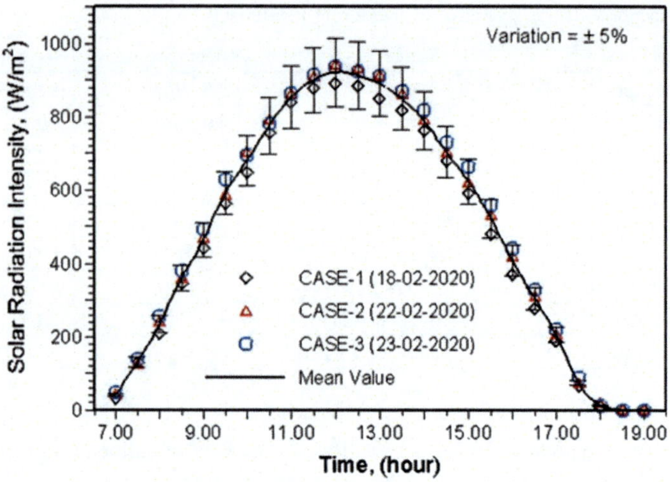


**Fig. 3** Variation of solar radiation intensity for all cases

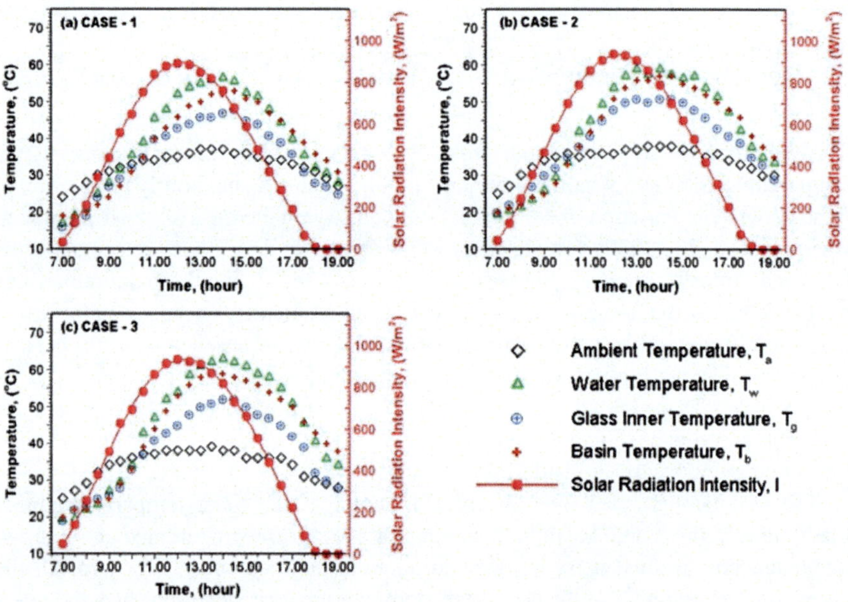


**Fig. 4** Variation of temperatures inside the solar still for different cases

conductivity that absorbs and transfers the heat quickly along with enhancement in the heat transferring area. That is the reason for higher difference in water temperature and glass temperature for CASE-3 and CASE-2 than that of CASE-1. Thus, the evaporation and condensation rates for CASE-3 are greater than others and thereby leading to increase in the freshwater yield for CASE-3 compared to CASE-2 and CASE-1.

Figure 5 shows the production flux for all the studied cases. Initially, the quantities of fresh water yield are almost similar, but as time progresses, the output of PSS with copper tubes rises at a higher rate as compared to solar still without copper tubes. However, the output distillate yield begins to drop in afternoon, after 3:00 PM. In evening, at 5:00 PM, the solar still with seven copper tubes (CASE-3) produces the highest distilled water than the others.

It is to be noted that the total amount of fresh water produced by CASE-3 is the highest with a value of 1.95 $L/m^2$ and subsequently followed by CASE-2 and CASE-1 with values of 1.62 $L/m^2$ and 1.22 $L/m^2$, respectively (see Fig. 6). It has been found that the distilled water output for CASE-3 is 60% higher than that of CASE-1 and 6% higher than CASE-2, while the CASE-2 output is 33% higher than that of the CASE-1. Enhancement in heat transferring area because of the introduction of copper tubes advances the heat transfer rate, which results in high water temperature and correspondingly rise in evaporation rate and thereby leading to the highest freshwater yield for CASE-3. The total dissolved salts (TDS), pH, and electrical conductivity ($dSm^{-1}$) of the produced fresh water are 28 mg/L, 7.02, and 0.08 $dSm^{-1}$, respectively, which are well within the range as per world health organization (WHO). Thus, the water produced from PSS is suitable for drinking purpose.

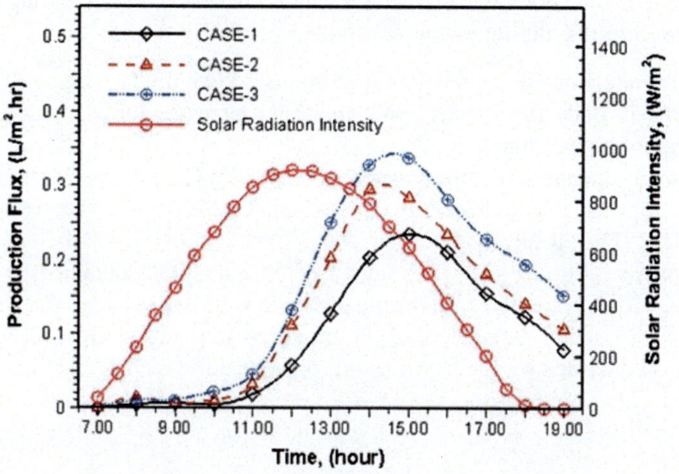


**Fig. 5** Comparison of production flux for different cases

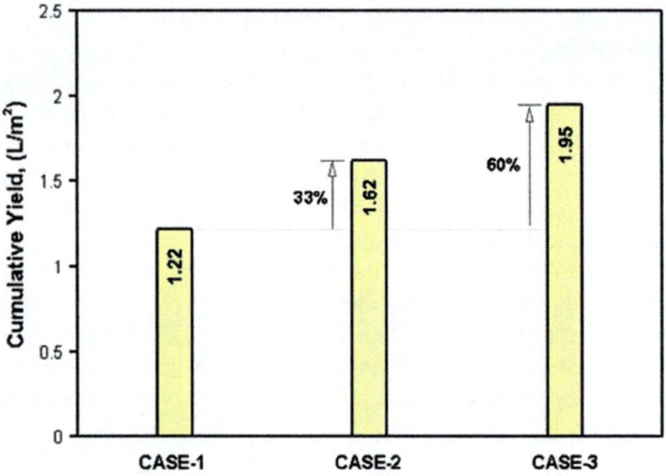


**Fig. 6** Productivity comparison of solar still with and without copper tubes

# 4 Conclusions

The influence of the placing copper tubes over the absorber of a PSS on the solar still performance is investigated experimentally in terms of fresh water productivity under the climatic conditions of Kanchikacherla, Andhra Pradesh, during February 2020. From results, the following concluding points are drawn:

- The introduction of copper tubes on the absorber of the solar still leads to increase in daily fresh water production and increases further with increasing number of copper tubes.
- The total amount of distilled water produced by CASE-3 is the highest, i.e., 1.95 L/m$^2$, which is followed by CASE-2 and CASE-1 with values of 1.62 L/m$^2$ and 1.22 L/m$^2$, respectively.
- Improvement in distilled water output of 60% and 33% obtained by CASE-3 and CASE-2, respectively, as compared with CASE-1.
- The water quality parameters of the produced fresh water are well within the range as provided by the world health organization (WHO).

# References

1. Sharma N, Varun, Siddhartha (2012) Stochastic techniques used for optimization in solar systems: a review. Renew Sustain Energy Rev 16(3):1399–1411
2. Sunil, Varun, Sharma N (2013) Modelling the drying kinetics of green peas in a solar dryer and under open sun. Int J Energy Environ 4(4):663–676

3. Nayi KH, Modi KV (2018) Pyramid solar still: a comprehensive review. Renew Sustain Energy Rev 81:136–148
4. Al-harahsheh M, Abu-Arabi M, Mousa H, Alzghoul Z (2018) Solar desalination using solar still enhanced by external solar collector and PCM. Appl Therm Eng 128:1030–1040
5. Khawaji AD, Kutubkhanah IK, Wie JM (2008) Advances in seawater desalination tech-nologies. Desalination 221(1–3):47–69
6. Taghvaei H, Taghvaei H, Jafarpur K, Estahbanati MK, Feilizadeh M, Feilizadeh M, Ardekani AS (2014) A thorough investigation of the effects of water depth on the performance of active solar stills. Desalination 347:77–85
7. Fathy M, Hassan H, Ahmed MS (2018) Experimental study on the effect of coupling parabolic trough collector with double slope solar still on its performance. Sol Energy 163:54–61
8. Yousef MS, Hassan H (2019) An experimental work on the performance of single slope solar still incorporated with latent heat storage system in hot climate conditions. J Cleaner Prod 209:1396–1410
9. Sakthivel TG, Arjunan TV (2019) Thermodynamic performance comparison of single slope solar stills with and without cotton cloth energy storage medium. J Therm Anal Calorim 137 (1):351–360
10. Arunkumar T, Murugesan D, Raj K, Denkenberger D, Viswanathan C, Rufuss DDW, Velraj R (2019) Effect of nano-coated CuO absorbers with PVA sponges in solar water desalting system. Appl Therm Eng 148:1416–1424
11. Panchal H, Hishan SS, Rahim R, Sadasivuni KK (2020) Solar still with evacuated tubes and calcium stones to enhance the yield: an experimental investigation. Process Saf Environ Prot 142:150–155
12. Bataineh KM, Abbas MA (2020) Performance analysis of solar still integrated with internal reflectors and fins. Sol Energy 205:22–36
13. Modi KV, Modi JG (2019) Performance of single-slope double-basin solar stills with small pile of wick materials. Appl Therm Eng 149:723–730
14. Jani HK, Modi KV (2019) Experimental performance evaluation of single basin dual slope solar still with circular and square cross-sectional hollow fins. Sol Energy 179:186–194
15. Kabeel AE, El-Maghlany WM, Abdelgaied M, Abdel-Aziz MM (2020) Performance enhancement of pyramid-shaped solar stills using hollow circular fins and phase change materials. J Energy Storage 31:101610–1–10
16. Panchal H, Mevada D, Sadasivuni KK, Essa FA, Shanmugan S, Khalid M (2020) Experimental and water quality analysis of solar stills with vertical and inclined fins. Groundwater Sustain Dev 11:100410

# Interference Effect Study on Wind Pressure Distribution in Buildings Using Computational Fluid Dynamics

**Vigneshwaran Rajendran, S. Prabavathy, and L. Sobankumar**

**Abstract** Usually, in cities, high rise buildings are commonly clustered together in the form of groups. The wind movement around the group of buildings is significantly different from an isolated building. The aerodynamic interference effect between two tall buildings is very much important to improve the occupancy comfort around the buildings. This paper investigates the wind-induced interference effect between two tall square buildings, and comparisons are made by changing the distance between the building. The analytical study is done by using computational fluid dynamics (CFD) incorporated with ANSYS Fluent software. Numerical simulation is carried out with a wind velocity of 10 m/s. Turbulence model *k*-ε is chosen for external flow simulation, because of its faster rate of convergence. From the analysis, it is evidence that for angular spacing of interference building with setback (AIBS), there is a reduction in pressure coefficient ($C_p$) by 13, 10, 8, and 9% on faces A, B, C, and D when compared with interference building with setback (IBS).

**Keywords** High rise buildings · Interference effect · Computational fluid dynamics (CFD) · Turbulence *k*-ε · Wake regions

## 1 Introduction

The demand for tall buildings is increasing day-by-day due to the lack of land available in urban areas. In order to find the actual wind load on the structure, investigations have to be carried out either analytically or experimentally. An experimental study is done with the help of wind tunnel testing but carrying a wind tunnel test is not economical, and it is also a time-consuming process. In order to avoid these difficulties, an analytical study is carried out with the help of ANSYS Fluent software to investigate the wind effect on tall buildings [1, 2]. Many previous research works have been carried out on different plan shapes in tall buildings such as L, T, E, C and Y plan shapes, and aerodynamic modifications are done in order

V. Rajendran (✉) · S. Prabavathy · L. Sobankumar
Mepco Schlenk Engineering College, Sivakasi, India



R. Kumar et al. (eds.), *Recent Trends in Thermal Engineering*, Lecture Notes in Mechanical Engineering, https://doi.org/10.1007/978-981-16-3132-0_10

to reduce the wind force acting on tall buildings [3, 4]. But nowadays, the study of interference effects is essential in order to find the actual wind behaviour on buildings. The impact of interference effects will lead to aerodynamic modifications. The factors affecting the interference excitations of tall buildings are size, shape and dynamic properties such as wind velocity, directions, terrain type and location. Many wind engineering codes such as ASCE 7-10, IS-875 (Part 3) 1987, AS/NZS:1170.2:2002 does not have any guidelines for accounting the interference effects in structural design [5, 6]. The author [7] investigated the interference effect on an isolated roughness flow for an isolated building, skimming flow for closed spacing buildings and wake interference flow for medium spacing buildings. The present investigation is focused on the interference effect between two tall buildings with the setback. Turbulence model RNG $k$-$\varepsilon$ is used for external flow simulation. The analysis is carried out for a 0° wind angle. Two square tall buildings with setback were chosen for the study. One is of interference building with setback (IBS) and the other is of angular spacing of interference building with setback (AIBS). For both the cases, building-1 represents principal building and building-2 represents interference building. Table 1 shows the spacing between the two buildings in the cases 2 and 3.

**Table 1** Details of the building model

| Building Models | Elevation | Plan |
|---|---|---|
| 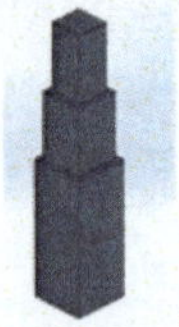 Case 1 - Single Building with Setback (SBS) | 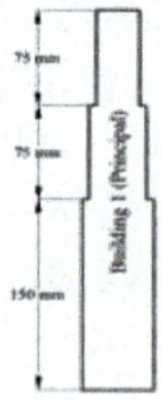 | 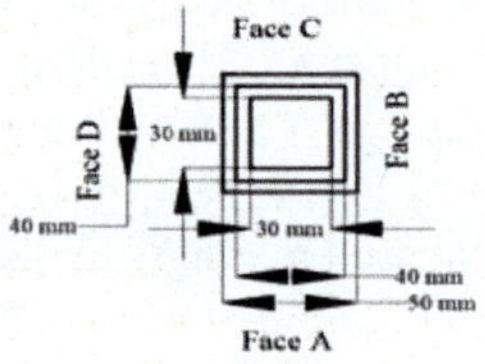 |
|  Case 2 - Inteference building with setback (IBS) | 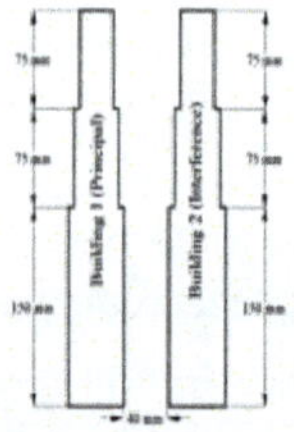 | 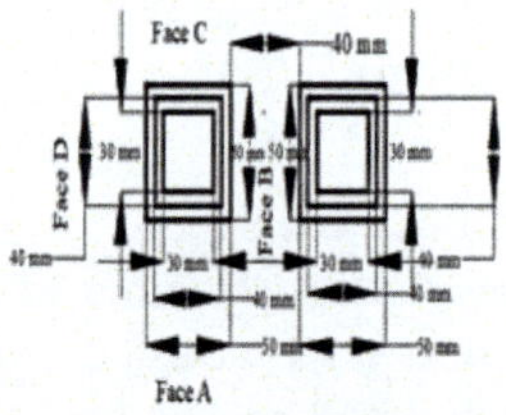 |
|  Case 3 – Angular spacing of Intereference building with setback (AIBS) | 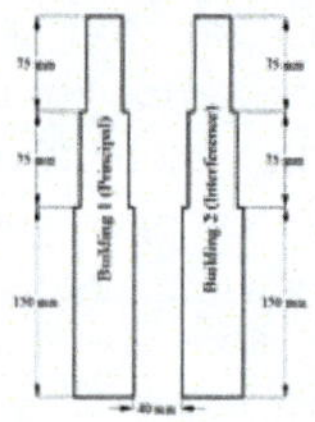 | 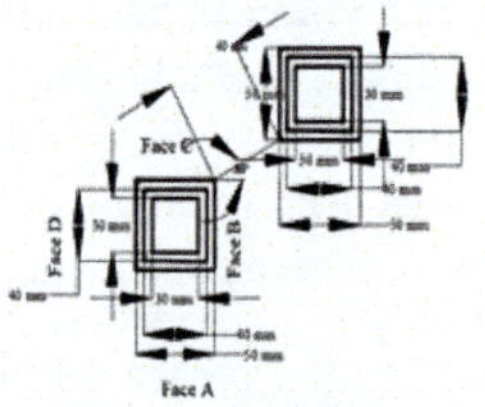 |

## 2 Analytical Methods

The numerical study is done using RNG $k$-ε turbulence model available in ANSYS Fluent package.

The transport equation momentum ($k$) and energy dissipation (ε) is given Eqs. 1 and 2.

$$\frac{\partial(\rho k)}{\partial t} + \frac{\partial}{\partial x_i}(\rho U_j k) = \frac{\partial}{\partial x_i}\left[\left(\mu + \frac{\mu_t}{\sigma_k}\right)\frac{\partial \varepsilon}{\partial x_i}\right] + P_k - \rho\varepsilon + P_{kb} \quad (1)$$

$$\frac{\partial(\rho \varepsilon)}{\partial t} + \frac{\partial}{\partial x_i}(\rho U_j \varepsilon) = \frac{\partial}{\partial x_i}\left[\left(\mu + \frac{\mu_t}{\sigma_\varepsilon}\right)\frac{\partial \varepsilon}{\partial x_i}\right] + C_1\frac{\varepsilon}{k}(P_k + C_3 P_b) - C_2\rho\frac{\varepsilon^2}{k} + S_\varepsilon \quad (2)$$

where $P_k$ is the mean velocity Shear, $P_b$ is due to buoyancy, $S_k$ is a user-defined. The values of the turbulence constants are $\sigma_k$ is 1, $\sigma_\varepsilon$ is 1.3, $C_1$ is 1.55, $C_2$ is 2.0, $C_\mu$ is 0.09 are used. For high Reynolds number, the energy dissipation may be assumed to $\rho k^{1.5}/l$.

$$\mu_t = \frac{C_\mu \rho l k^2}{\varepsilon} \quad (3)$$

where $\mu_t$ and $k$ is nonzero, $\rho$ is the fluid density, $C_\mu$ is a constant of value 0.09.

## 3 Details of the Model

The details of the building model is represented in Table 1.

### 3.1 Domain Size, Meshing and Grid Indepedency Study

The computational domain is build based on the recommendation given by frankle. The domain is built-in the upstream side 5H, the side faces both left and right 5H and in the top face 5H, where H represents H of the building model. Further, the downstream faces 15H is provided in order to capture the flow behaviour. Figure 1 represents the computational domain size, Fig. 1a 3D representation of the computational domain, Fig. 1b side view of the domain, Fig. 1c front view of the computational domain. No-slip is considered for the building face, and free slip is given for top and side face of the domain. The computational domain is operated with 1 atm pressure, i.e. 101,325 Pa. Reynolds number of the model varies from $3.7 \times 10^6$ to $4.0 \times 10^6$.

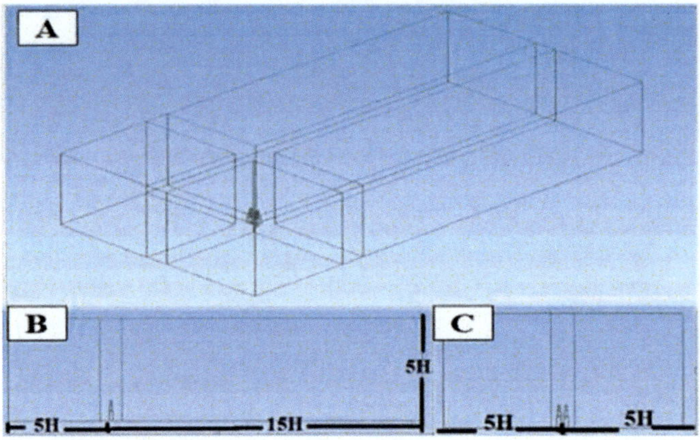


**Fig. 1** Size of computation domain **a** 3D view, **b** side view, **c** front view

Meshing is very much important in order to decide the quality of the results. The size of the mesh influences the outcome of the pressure. Very fine mesh is adopted for the building faces when compared to the computational domain. Figure 2a shows the mesh for the computational domain 2b shows the mesh pattern for single building with setback (SBS), and Fig. 2c shows the mesh pattern for the interference building with setback (IBS). Figure 2d shows the mesh pattern for angular spacing of interference building with setback (AIBS).

The grid independency test is performed in order to check the quality of the results. The meshing is done in four categories large, coarse, medium, and fine grids. The variation of $C_p$ with respect to the mesh is shown in Table 2. The fine mesh is done till the final $C_p$ reaches to zero percentage error. The wall distance $\Delta_x > K_s$. For the current study, the value of $y+$ is 30, and the distance of the first cell near the building is 0.002 m.

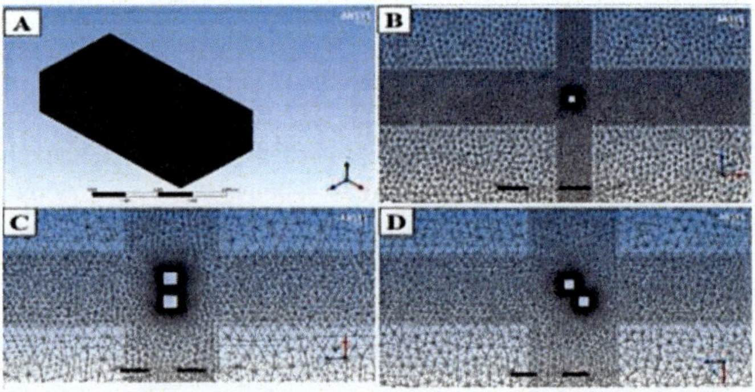


**Fig. 2** Meshing **a** 3D view, **b** isolated building with setback, **c** interference building with setback, **d** angular spacing of Interference building with setback

**Table 2** Components of computational grids

| | Name | $\Delta x$ (m) | No of node | No. of elements | Pressure coefficient ($C_p$) | % of error |
|---|---|---|---|---|---|---|
| Case 1 | Course | 0.003 | 34,375 | 1,87,935 | 0.40 | 10 |
| | Basic | 0.0025 | 78,583 | 4,19,335 | 0.44 | 10 |
| | Fine | 0.002 | 3,69,602 | 20,50,983 | 0.49 | 11.365% |
| | Very fine | 0.002 | 1542762 | 88,13,077 | 0.49 | – |
| | Course | 0.003 | 2,77,972 | 10,18,856 | 0.48 | 4.16% |
| Case 2 | Basic | 0.0025 | 3,22,078 | 12,49,451 | 0.50 | 4.1% |
| | Fine | 0.002 | 6,11,655 | 28,77,797 | 0.51 | 0.5% |
| | Very fine | 0.002 | 17,85,924 | 96,40,608 | 0.51 | – |
| | Course | 0.003 | 3,23,494 | 10,18,856 | 0.33 | 6.06% |
| Case 3 | Basic | 0.0025 | 3,64,596 | 12,49,451 | 0.35 | 11.42% |
| | Fine | 0.002 | 6,46,257 | 28,77,797 | 0.39 | 6.1% |
| | Very fine | 0.002 | 17,85,924 | 96,40,608 | 0.39 | – |

### 3.2 Solver Settings and Atmospheric Boundary Layer (ABL)

A user-defined ABL is used for simulating the wind profile. The simulation is carried using open terrain condition. A steady-state simulation is performed using RANS equation in built-in ANSYS Fluent software. A SIMPLE algorithm is used for coupling pressure and velocity. The iterations were performed till the variables *X*, *Y* and *Z* momentum direction and continuity *k* and ε are achieved when the residual reaches a value of $1 \times 10^{-4}$. Atmospheric boundary layer profile is generated using Eq. (4)

$$\frac{U}{U_0} = \left(\frac{Z}{Z_0}\right)^{\alpha} \tag{4}$$

where $U_0$ is the wind speed taken as 10 m/s, *U* is the velocity at a particular height (*Z*), and $Z_0$ is the boundary layer height.

### 3.3 Validation

Validation has been done on a square shape building of plan size, 100 mm × 100 mm and building height 500 mm. Using RNG *k*-ε turbulence model and the CFD analysis is carried for a uniform flow with a wind velocity of 10 m/s is provided at the inlet. From the result, it is observed that the package is approximately close to face average value of the coefficient of pressure mentioned in the IS:875

**Table 3** Validation of pressure coefficient ($C_p$) with different wind engineering codes

| Wind loading code | Face A ($C_p$) | Face B ($C_p$) | Face C ($C_p$) | Face D ($C_p$) |
|---|---|---|---|---|
| ANSYS Fluent 19.2 | 0.90 | −0.20 | −0.86 | −0.85 |
| IS:875(Part 3) 2015 | 0.80 | −0.25 | −0.80 | −0.80 |

(Part 3) 2015. Table 3 displays the value of the pressure coefficient ($C_p$) for square building and the comparison between IS:875 (Part 3):2015 wind Engineering codes.

## 4 Result and Discussion

The study on the interference effects using CFD is carried out using the turbulence model RNG $k$-ε. The mean pressure coefficient on each face of the buildings is a non-dimensional parameter and is calculated using Eq. 5.

$$C_{p,\mathrm{mean}} = \frac{p - p_0}{\frac{1}{2}\rho U_H^2} \tag{5}$$

where $p$ is the pressure from the point, $p_0$ is the pressure at reference height, $\rho$ is the density of the air is 1.225 kg/m$^3$. $U_H$ is the mean wind velocity at the height $H$.

### *4.1 Pressure Distribution on Single Building with Setback (SBS): Case 1*

The setback building is divided into three segments from the base. From 0 to 15 cm face A1, 15 to 22.5 cm face A2, 22.5 to 30 cm face A3, respectively. Corresponding identification is given to the side faces and leeward faces of the building. Figure 3 shows the pressure coefficient ($C_p$) contour on the various face of the IBS. For case 1, i.e. IBS faces A1, A2 and A3 experiences positive pressure distribution, due to windward direction, whereas the other face B1, B2, B3, C1, C2, C3, D1, D2 and D3 experiences negative $C_p$, because these building faces lies in leeward direction. For the windward faces A1, A2 and A3, high positive pressure is highly concentrated in the middle position of the building faces and the pressure decrease along the edges as shown in Fig. 3a. In further observation, the top face A3 the pressure decreases along the top due to the dispersion of the flow on the top. On the other hand, for the faces C1, C2, C3 experiences high negative ($C_p$) because it lies in the leeward direction, and the pressure increase along the edges in the top of the building face C2 and C3. However, the comparatively less negative $C_p$ is observed in the side faces B1, B2, B3, D1, D2 and D3. Figure 4 shows the

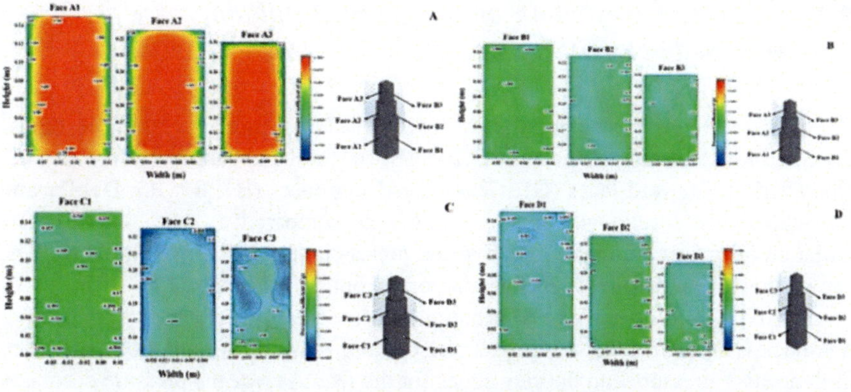


**Fig. 3** Pressure contour for the case 1 on each face of the building with setback **a** building faces A1, A2, A3, **b** building faces B1, B2, B3, **c** building faces C1, C2, C3, **d** building faces D1, D2, D3

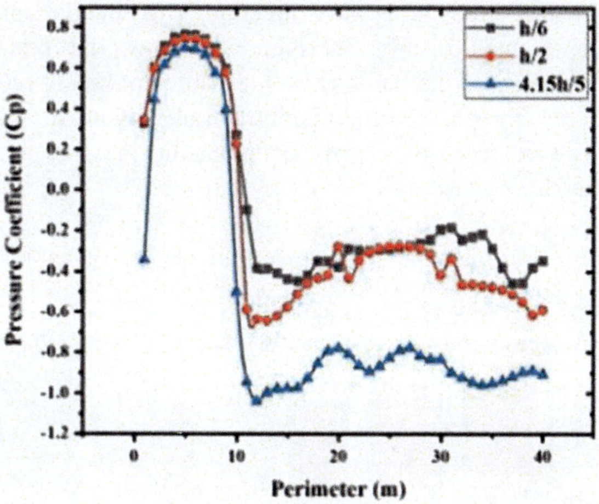


**Fig. 4** Pressure coefficient ($C_p$) measure along the perimeter of the building at heights h/6, h/2 and 4.15 h/5 for the case 1

measure of the pressure coefficient along the perimeter of the building at three different levels h/6, h/2 and 4.15 h/5 for the case SBS. The pressure coefficient ($C_p$) in the windward faces experiences similar contour in three different height h/6, h/2 and 4.15 h/5. However, in the leeward faces, mild variation in pressure coefficient is observed at the height h/6, h/2, and the greater difference is observed in the top height 4.15 h/5 and undergoes high negative pressure coefficient ($C_p$).

## 4.2 Pressure Distribution on Inteference Building with Setback (IBS): Case 2

For the case 2 interference building with setback (IBS), similar contour pressure is observed for this case as that of case 1 (SBS) for the windward faces A1, A2, A3. However, the leeward faces (C1, C2, C3) and side faces (B1, B2, B3, D1, D2 and D3) experience a mild variation in contour as compared to case 1 (SBS). The variation in pressure contour is due to the presence of neighbouring building. The shows that the presence of the neighbouring building can result in a change of pressure contour. The pressure distribution for the case 2, interference building with a setback is shown in Fig. 5, the pressure contour in the windward face experience high positive pressure and the side edges for the face A1, A2 the pressure decreases because of the dispersion of the flow in the side face. However, the pressure at the top position in the face A3, the pressure reduces at the top and side faces due to the dispersion of the flow in three directions. For the side faces B3 as shown in Fig. 5 experiences very low-pressure coefficient ($C_p$) due to the presence of the neighbouring building, whereas in the side face D3 the positive $C_p$ is observed because there is no adjacent building. Faces C1, C2 and C3 experience very low $C_p$ value because it lies in leeward direction. This low pressure increases along the edges as represented in Fig. 5c. Figure 6 shows the pressure measurement along the perimeter of the building at three different height h/6, h/2 and 4.15 h/6. Windward faces experience equal distribution of pressure at a different height, whereas in the leeward faces experience a fluctuating pressure at all the levels of the building height.

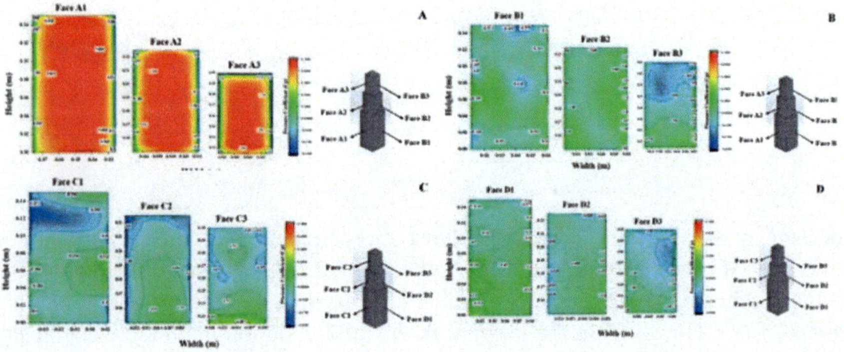

**Fig. 5** Pressure contour for the case 2 on each face of the building with setback **a** building faces A1, A2, A3, **b** building faces B1, B2, B3, **c** building faces C1, C2, C3, **d** building faces D1, D2, D3

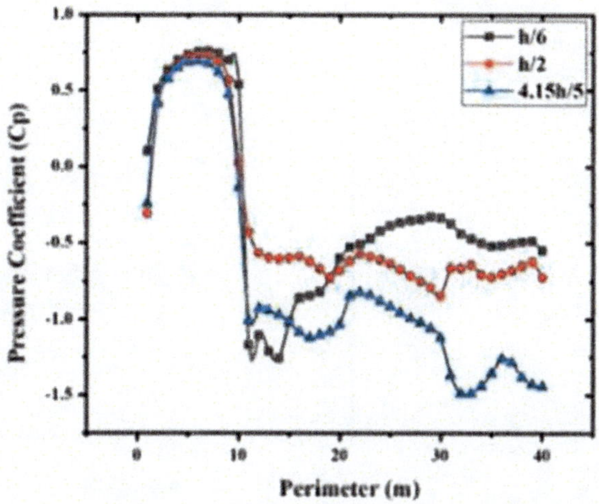


**Fig. 6** Pressure coefficient ($C_p$) measure along the perimeter of the building at heights h/6, h/2 and 4.15 h/5 for the case 2

## 4.3 *Pressure Distribution on Angular Spacing of Interference Building with Setback (AIBS): Case 3*

For the case 3, the angular spacing of interference building with a setback, building 1, i.e. principal building experiences positive $C_p$ in the windward faces, whereas for the leeward faces C1, C2, C3 experiences negative pressure ($C_p$). Figure 7 shows the pressure contour on the various face in the principal building for case 3. The top side faces B3 experiences high negative $C_p$ and further negative $C_p$ is observed in the leeward faces C2 and C3, respectively. Figure 8 shows the pressure coefficient

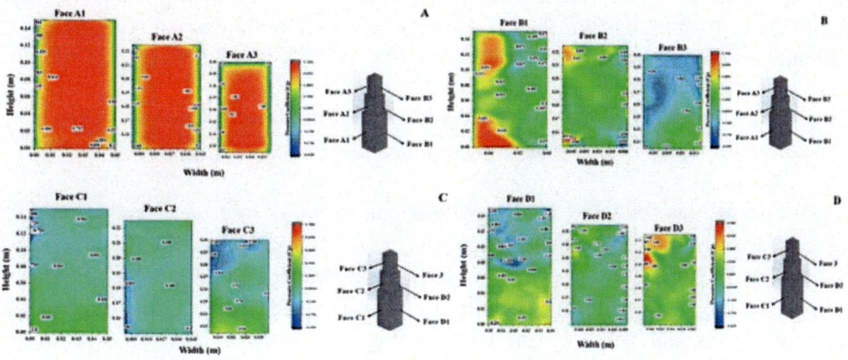

**Fig. 7** Pressure contour for the case 3 on each face of the building with setback **a** building faces A1, A2, A3, **b** building faces B1, B2, B3, **c** building faces C1, C2, C3, **d** building faces D1, D2, D3

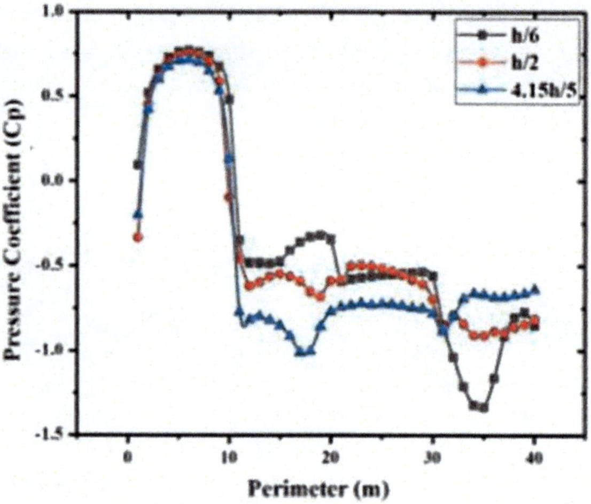


**Fig. 8** Pressure coefficient ($C_p$) measure along the perimeter of the building at heights h/6, h/2 and 4.15 h/5 for the case 3

($C_p$) measured along the perimeter at different heights h/6, h/2 and 4.15 h/5. Positive faces experience the same pressure coefficient at all the levels, whereas in the negative face experiences different pressure contour in all the three levels.

## 4.4 Interference Effects

Interference effect arises due to the presence of adjacent buildings and largely depends upon the terrain category and magnitude of wind force. Interference effect is represented by non-dimensional factors as aerodynamic forces acting on the building-1 because of the adjacent building-2. In the present study, the buildings are subjected to boundary layer wind flow at a wind angle 0°. For the first case, the distance between the building-1 (Principal building) and building-2 (Interference building) is 40 mm, and for the second case, both the buildings are separated by a distance of 40 mm at an angle of 30°.

$$I.F = \frac{\text{Mean pressure for a face in interference condition}}{\text{Mean pressure for a face in insolated condition}}$$

The guidelines for wind load modifications in the design of a square building with adjacent square buildings. $C_{p,\ \text{interfering}} = I.F \times C_{p,\ \text{isolated}}$. Interference effect is calculated from the face mean pressure of the building with interference condition to the face mean pressure of the building when isolated [8, 9]. Table 4 shows the pressure coefficient for different faces. Table 5 shows the $C_{p,\ \text{interference}}$ values obtained in the principal building-1 for case 2 and 3. From the result, it is observed that the value of the $C_p$ for the case 2 (IBS) for the faces A, C and D, and for the case 3 (AIBS) due to angular spacing of the interference building, there is no much

**Table 4** Pressure coefficient ($C_p$) for different cases

| Building faces | Pressure coefficient ($C_p$) Case 1—Single building with setback (SBS) | Pressure coefficient ($C_p$) Case 2—Inteference building with setback (IBS) | Pressure coefficient ($C_p$) Case—3 Angular spacing of intereference building with setback (AIBS) |
|---|---|---|---|
| Face A | 0.49 | 0.51 | 0.39 |
| Face B | −0.65 | −0.52 | −0.31 |
| Face C | −0.8 | −0.76 | −0.67 |
| Face D | −0.68 | −0.55 | −0.4 |

**Table 5** Interference pressure coefficient $C_{p,\ int}$ value for IBS and AIBS

| Building—1 Faces | IBS $C_{p,\ \text{inter}}$ ($C_{p,\ \text{interfering}} = I.F \times C_{p,\ \text{isolated}}$) | AIBS $C_{p,\text{inter}}$ ($C_{p},\ _{\text{interfering}} = I.F \times C_{p,\ \text{isolated}}$) |
|---|---|---|
| Face A | 0.51 | 0.39 |
| Face B | −0.52 | −0.31 |
| Face C | −0.61 | −0.67 |
| Face D | −0.44 | −0.4 |

variations in pressure on all the four faces (A, B, C and D). This is because due to the consideration of wind effects in adjacent buildings, the pressure obtained in the principal building varies.

## 5 Conclusion

A comprehensive study is made on the interference effect between two buildings with the setback. The analysis is carried out for 0° wind angle for all the three cases, and the following are the conclusions made from this study.

The wind flow around an isolated building is different from the grouped building. Wind flow pattern has a direct impact on pressure coefficient and leads to either increase or decrease in pressure coefficient. Grid sensitive analysis is performed in order to obtain the accuracy of the results, by using ANSYS Fluent software package. Turbulence model $k$-$\varepsilon$ is very much useful in capturing the pressure distributions on buildings, and it is widely adopted for external flow problems. The wind pressure on the building varies with respect to the presence of neighbourhood buildings, in the present study for case 3 (AIBS), there is 30.6% reduction in interference pressure coefficient ($C_{p},\ _{\text{int}}$) in face A, due to the presence of angular spacing followed by leeward faces B by 40.38%, C by 11.8% and D by 37.5% when compared with IBS. This evidences that the pressure coefficient ($C_p$) on each face of the building depends on the presence of neighbouring buildings. For all the cases, the positive $C_p$ (windward faces) decreases on the edges of the buildings due to the dispersion of the flow, and the negative $C_p$ (leeward faces) increases on the edges of the buildings due to the convergence of the flow. With the help of CFD analysis, it is very much useful to capture the pressure distribution on each face of the building, and the above study is validated with both Indian and American code for the design of wind loads on buildings.

**Acknowledgements** The author would like to thanks to the department of Civil Engineering in Mepco Schlenk Engineering College (Autonomous), Sivakasi, for providing adequate facilities to carry out the study and Institute of Engineers (India) IE (I) for its grateful financial support (Project ID: DR2020004) to complete the project work successfully.

## References

1. Chen B, Shang L, Qin M, Chen Z, Yang Q (2018) Wind induced effects of high-rise building on low-rise building with flat roof. J Wind Eng Ind Aerodyn 183:88–113
2. Kim W, Tamura Y, Yoshida A (2015) Interference effects on aerodynamic wind forces between two building. J Wind Eng Ind Aerodyn 147:186–201
3. Chakraborty S, Dalui SK, Ahuja AK (2014) Wind load on irregular plan shape tall building—a case study. Wind Struct 19(1):59–73
4. Mukherjee S, Chakraborty S, Dalui SK, Ahuja AK (2014) Wind induced pressure on 'Y' plan shape tall building. Wind Struct 19:523–540
5. Sy LD, Yamada H, Katsuchi H (2019) Interference effects of wind-over-top flow on high-rise buildings. J Wind Eng Ind Aerodyn 187:85–96
6. Yu X, Xie Z, Gu M (2018) Interference effects between two tall buildings with different section sizes on wind-induced acceleration. J Wind Eng Ind Aerodyn 182:16–26
7. Lo YL, Kim YC, Li YC (2016) Downstream interference effect of high-rise buildings under turbulent boundary layer flow. J Wind Eng Ind Aerodyn 159:19–35
8. Hui Y, Tamura Y, Yang Q (2017) Analysis of interference effects on torsional moment between two high rise buildings based on pressure and flow field measurement. J Wind Eng Ind Aerodyn 164:54–68
9. Lo YL, Kim YC, Yoshida A (2017) Effects of aerodynamic modification mechanism on interference from neighboring buildings. J Wind Eng Ind Aerodyn 168:271–287

# Design of PID Controller to Reduce the Vibrations of a Vertical Washing Machine

Harbhinder Singh, Munish Mehta, and Pramod Kumar

**Abstract** This paper focuses on a vertical load washing machine system for vibration analysis and on its different parameters used to control the vibration intensity. Vibration control basically depends upon two major factors, i.e. vibration absorber and vibration isolator. Vibration isolation may be a common method to protect receivers (device, human, system, machine or structure) from the vibration source. For this purpose, a virtual model was designed using MATLAB and Simulink tool. After creating the model, the analysis of vibration was performed. PID controller was used to find the optimum value of controller parameters like $K_P$, $K_I$, $K_D$, for obtaining a stable as well as responsive system. The observed results will help the designers in designing the machine efficiently by using the knowledge of the effects of different parameters.

**Keywords** Vertical washing machine · Unbalanced load · PID controller · MATLAB

## 1 Introduction

Nowadays, the demand for energy saving is increasing rapidly. Energy saving demand is important in industrial applications and home appliances such as vacuum cleaners, washing machines, ventilation and heating systems. Thus, there is a need of solutions which are not only cost-effective but also these do not make compromises on the quality of the products. A washing machine is very complicated system with combination of dynamics and electronics control system. Washing machine is classified into two categories: top-loading, i.e. vertical axis washer, and frontloading, i.e. horizontal axis washer. Top-loading washing machine has four suspension rods through which wash unit is suspended. For washing machine

H. Singh · P. Kumar
U.I.E.T, Panjab University, Chandigarh, India

M. Mehta (✉)
School of Mechanical Engineering, Lovely Professional University, Jalandhar, Punjab, India

 
R. Kumar et al. (eds.), *Recent Trends in Thermal Engineering*, Lecture Notes in Mechanical Engineering, https://doi.org/10.1007/978-981-16-3132-0_11

having one non-rotating washtub with one rotating basket, Spelta [1] proved that damping control ensures a low level of vibration for every working conditions of the washing machine. The power is transmitted to the identical rotating axis of the basket through a belt from a motor during both the washing and spin-drying. After the washing mode, the laundry is dried by the spin-drying mode. During the washing mode, Phule [2] described that clothes tend to clump and create an unbalance mass in the basket. As a result, the unbalanced mass causes serious vibration problems when the basket spins with relatively high speed during the drying cycle. Larger the spinning speed lower the drying time and higher the centrifugal force produced by unbalanced mass. Tsujiuchi [3] designed and developed a new suspension system for vertical axis washer with CREO and analysis tool ADAMS. In this study, different parameters were rotational speed, eccentricities of the centre of rotation and hydraulic force. Due to centrifugal action clothes pressed against the inner wall of the spinning drum, and this can exert a large unbalance force on washing machine wall. Weight of the laundry and mode of washing [4] decides the amount of unbalanced mass. The industrial focus is on increasing efficiency of a washer, low vibration and noise. So, these areas getting more attention and have been actively studied. The objectionable effect of washer working is vibration and noise. Poor efficiency means large energy consumption, noise and vibration. In the research study, Narkhede [5] concluded that the proposed damping control and vibration levels were low for working conditions. Unwanted vibration can be reduced by active vibration control. Active vibration technique uses different kinds of sensors to measure the motion or force or acceleration of the vibrating mechanical system. A powered actuator is used to generate a force to resist the unwanted motion. Papadopoulos [6] worked on a method of assigning property values to increase isolation efficiency for optimum design. PID controller works on the basis of how much steady-state error is there and fix it to the desired value with the help of proportionality, integration and derivation. For reducing the transient horizontal vibration of top-loading washing machine, Kim [7] introduced, a quad horizontal damper system, a new vibration reduction technology and analysed the vibration characteristics using Newtonian mechanics. Kumar [8] designed a model of portable washing machine considering the parameters of cost-effective and three-dimensional horizontal axes. Instability of washing machine was predicted during the spinning cycle.

In this paper, a PID controller was proposed to minimize vibrations. To optimize the value of various parameters, PID tuner in MATLAB toolbox was used. PID controller is a closed-loop system, when input parameters change due to excitation force during wash cycle and spin cycle, corresponding signals travel to the system to change the input parameters according to reduce the vibrations. Numonoy [9] concluded that along with a variant speed of whirling, which has a more obvious influence than the lateral forces at a unique operational speed, unbalanced force depends on the speed of whirling.

# 2 Materials and Methods

Methodology of the work is as shown in Fig. 1. Vibrations in washing machines are produced due to unbalanced mass in drum. Large intensity of vibrations may lead to noise and reduce the work efficiency of the machine by damaging the machine components. If the unbalanced machine is used for washing for a long time, then, due to vibrations of unbalanced mass, the electronic components may fail.

## 2.1 *Washing Machine Selection*

Traditional top-loading models are the least efficient of all washer types. Top-loading washing machines consume more energy during drying time, per cycle fewer clothes are washed and large amounts of water are consumed. Also, intensity of vibration is bigger challenge for vertical axis washing machines. Therefore, this work has been carried out on a vertical washing machine.

## 2.2 *Mathematical Model Formulation*

The mechanical equation of motion, i.e. the dynamic balance of forces, is given by

$$M\ddot{X} + C\ddot{X} + KX = F_0 \sin \omega t \tag{1}$$

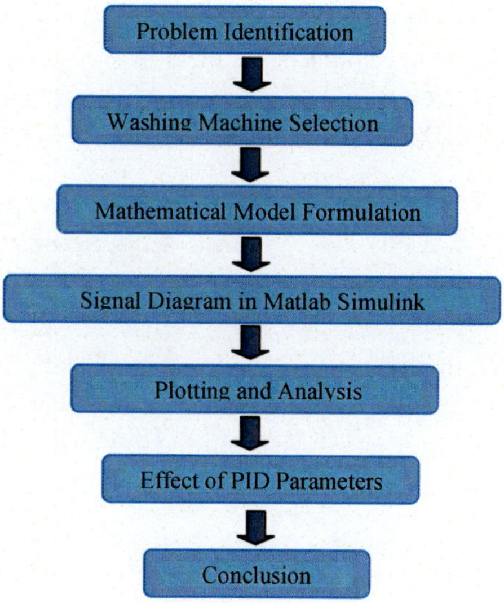


**Fig. 1** Flowchart of work methodology

The amplitude of study state vibration of the above model is given as:

$$X = \frac{m_0 e\omega^2/K}{\sqrt{\left[1 - \left(\frac{\omega}{\omega_n}\right)^2\right]^2 + (2\xi\frac{\omega}{\omega_n})^2}} \tag{2}$$

where

$$\xi = \frac{C}{2M\omega_n} \tag{3}$$

Nomenclature:
$\xi$ = Damping ratio
$m_0$ = Unbalance mass, kg
e = Eccentricity, mm
$\omega$ = Operating speed, rpm
K = Spring stiffness N/mm
$\omega_n$ = Natural frequency of washer, rpm
C = Coefficient of damping, N-s/m
The above equation has been used to form a mathematical model based on the parameters.

## 2.3 *Import the Value of Input Parameter*

The input parameters were selected to carry out the research, and data values of input parameters were imported into the simulation model. The parameters used in the simulation model are mass of the washing machine, mass of dry clothes, damping coefficient, stiffness of spring, etc., as shown in Table 1.

**Table 1** Considered values of parameters

| Name of input parameter | Symbol | Value | Units |
|---|---|---|---|
| Spring constant | K | 2250 | N/m |
| Damping constant | C | 300 | Ns/m |
| Mass of washing machine | M | 38 | Kg |
| Mass of load | M | 7 | Kg |
| Eccentricity | E | 0.25 | M |
| Speed | N | 720 | RPM |

### 2.4 Plot and Analyse the Graph of Amplitude

The next important step was to plot the values to analyse the data. The results obtained from models with PID and without PID were compared on the graph, and amount of reduction was calculated.

### 2.5 Effect of PID Parameters

Behaviour was plotted on a graph corresponding to each one of parameters in PID controller. The effect of PID parameter was prominent to reduce the vibrations in the system.

## 3 Modelling of PID Controller

Model shown in Fig. 2 (without PID) is the simulation of washing machine and to obtain the desired results; PID controller came into picture. To form a PID controller, a proportional gain named as $K_P$, an integrator with gain named as $K_I$ and a derivative block with gain named $K_D$ are needed. All three of these are connected in the manner as shown in Fig. 2.

After the formation of simulation model with PID as shown in Fig. 3, different values of $K_P$, $K_I$ and $K_D$ are randomly chosen, and then results obtained from scope block are recorded.

Figure 4 shows the combined model circuit with and without PID. The results obtained from model with PID and without PID controller are compared, and the reduction or incrimination after the application of PID is calculated. For optimality, different values for $K_P$ are considered by keeping the other two parameters, i.e. $K_I$ and $K_D$ constant. The results obtained are plotted on a graph describing the

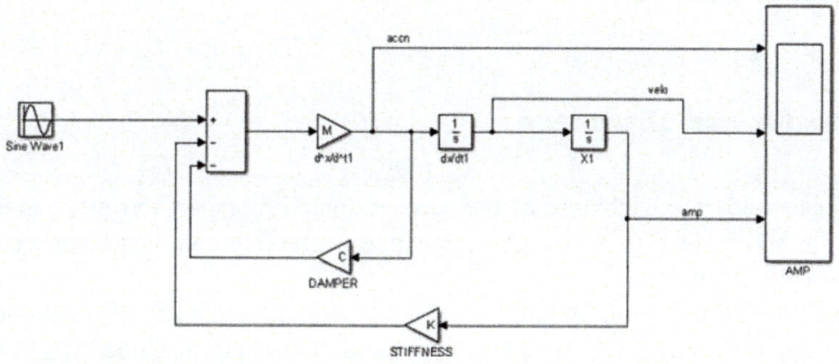


**Fig. 2** Simulink model without PID

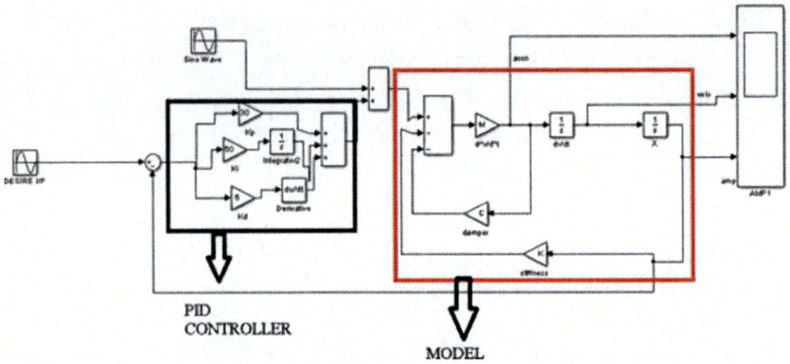


**Fig. 3** Simulation model with PID controller

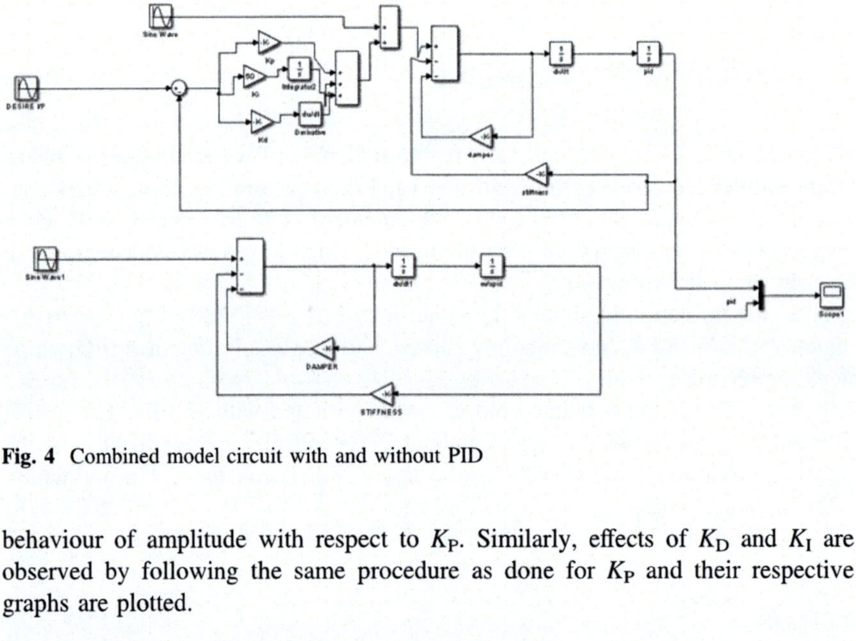

**Fig. 4** Combined model circuit with and without PID

behaviour of amplitude with respect to $K_P$. Similarly, effects of $K_D$ and $K_I$ are observed by following the same procedure as done for $K_P$ and their respective graphs are plotted.

## 4 Results and Discussion

In order to study the behaviour of the acceleration with respect to time, a graph is plotted as shown in Fig. 5 by using the experimental data that had been recorded using accelerometer.

After the formation of Simulink model of washing machine, as mentioned in Table 1, the input parameter values that are mass of washing machine (*M*), stiffness of spring (*K*), damping coefficient of damper (*C*) mass of unload/dry cloth (*m*),

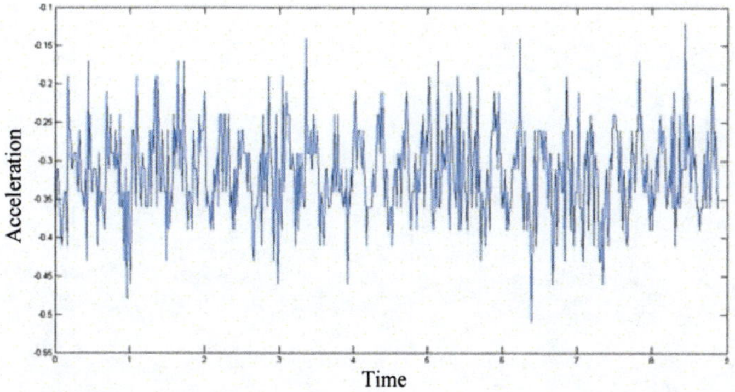


**Fig. 5** Plot of data recorded in accelerometer

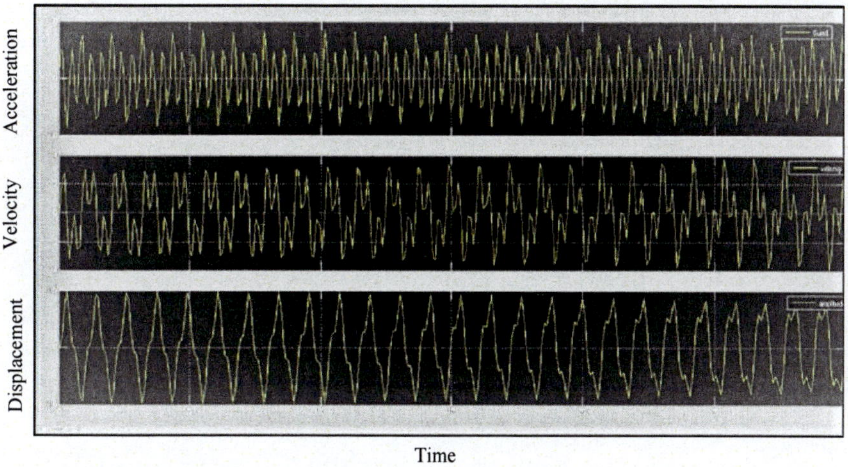


**Fig. 6** Plot of amplitude, velocity and acceleration against time

eccentricity ($e$) are utilized. After simulation, the output from scope block is obtained in the form of three graphs, that is, acceleration against time, velocity against time and amplitude against time, respectively, as shown in Fig. 6.

In order to reduce amplitude of vibration, a PID controller is used additionally as shown in Fig. 3 and values of $K_P$, $K_I$ and $K_D$ are provided randomly to check the behaviour of graph as shown in Fig. 7.

Euclidean norm was used to obtain the relative difference between the original value and PID controlled values. MATLAB code for Euclidean is shown in Fig. 8. The matrix A is defined as the square root of the sum of the absolute squares of its elements and isgiven as:

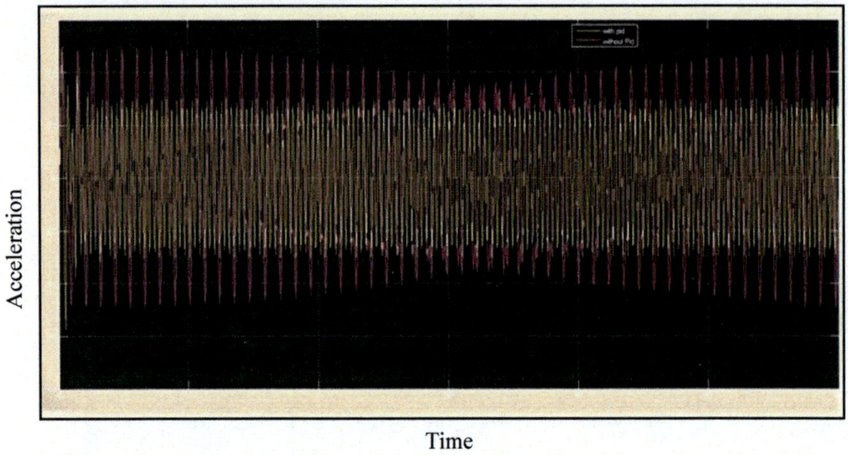


**Fig. 7** Amplitude comparison with and without PID

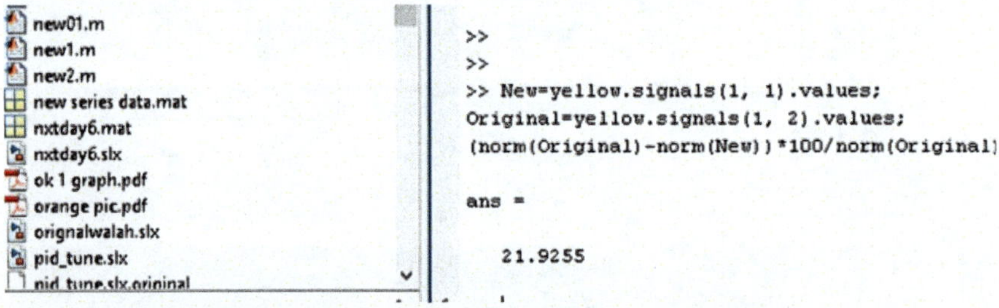


**Fig. 8** MATLAB code for Euclidean norm calculation

$$\|A\|_F \equiv \sqrt{\sum_{i=0}^{m}\sum_{j=1}^{n}|a_{ij}|^2} \tag{4}$$

And in this case where $K_P = 600$, $K_I = 150$, $K_D = 50$, we get the amplitude reduction in percentage as 2.36%.

Figure 9 shows the output of manually tried combination of values of $K_P = 500$ $K_i = 100$ and $K_D = 20$, and the amplitude reduction in percentage was 21.18%.

As seen from the above graph, the amplitude of vibrations is reduced. As the amplitude is reduced corresponding acceleration of vibrations is also reduced by reasonable amount. When compared with and without PID controller, the reduction is approximately 21%, which is satisfactory amount of reduction. Figure 10 shows the behaviour of $K_P$ varied from 0 to 2,000 with step-size of 100, and the values of $K_I$ and $K_D$ are kept constant at 100 and 20, respectively.

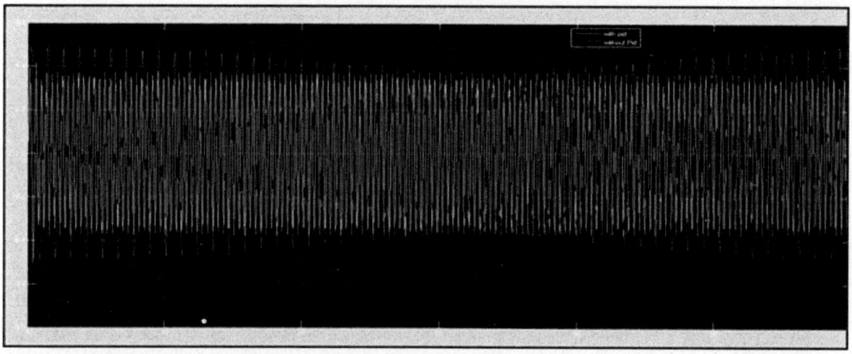


**Fig. 9** Amplitude at $K_P = 500$, $K_I = 100$ and $K_D = 20$

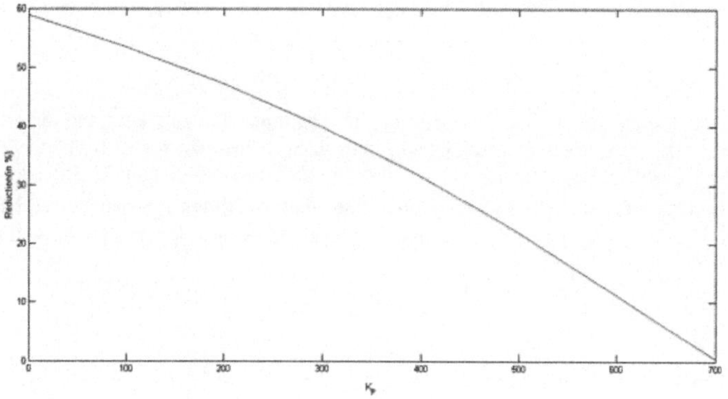


**Fig. 10** Effect of $K_P$ by keeping $K_I = 100$ and $K_D = 20$

## 5 Conclusions

On the basis of experimental results from MATLAB R2013a, it was observed virtually that mechanical vibrations of vertical axis washing machine can be effectively controlled using PID controller. In this research, PID controller was used successfully and amplitude of vibration was reduced by 21.18%. Further, the effect of changing the key components *P*, *I* and *D* in PID controller was obtained. Amount of reduction in amplitude of vibration was observed by changing the value of $K_P$, $K_D$ and $K_I$, separately and graphs were plotted. More controllers like fuzzy and neuro-fuzzy can be explored using simulation, and their effect on the vibrations can be observed. With the help of simulation, optimized value of different parameters *P*, *I* and *D* can be found for the desired amount of reduction in vibration and then it can be directly implemented empirically on the real machine.

## References

1. Spelta C, Previdi F, Savaresi SM, Fraternale G, Gaudiano N (2009) Mechatronics control of magnetorheological dampers for vibration reduction in a washing machine. Mechatronics 19 (3):410–421
2. Phalak MK, TA J, KV P (2018) Univeristy; Design and development of vertical axis washing machine suspension system for stability. Evol Mech Eng 2(1):1–8. https://doi.org/10.31031/EME.2018.02.000526
3. Tsujiuchi N, Koizumi T, Matsumoto S, Yamaoka N (2007) Vibration analysis and modeling of adrum type washing machine. Proc Conf KansaiBranch 2007(82):43–44
4. Shihab IM (2017) Theoretical and experimental study of the vibration of a drum type washing machine at different speeds Al-Nahrain. J Eng Sci 20(5):1160–1171
5. Narkhede CN, Dhande KK (2016) Vibration reduction of a vertical axis drum based washing machine. Imp J Interdiscip Res 9(2):794–802
6. Papadopoulos E, Papadimitriou I (2001) Modeling, design and control of a portable washing machine during the spinning cycle. In: Proceedings of IEEE/ASME international conference on advanced intelligent mechatronics systems, Como, Italy, pp 899–904
7. Kim YY, Kim DC (2019) Dynamic modelling and analysis of a quad horizontal damper system for transient vibration reduction in top loading washing machine. J Mech Sci Technol 33:1123–1130. https://doi.org/10.1007/s12206-019-0210-8
8. Kumar S, Singh P, Sehgal S, Kumar H, Aggarwal N, Singh S (2020) Application of industrial internet of things for online monitoring of bearings. Recent advances in mechanical engineering. Lecture notes in mechanical engineering, Springer, pp 333–342 https://doi.org/10.1007/978-981-15-1071-7_28
9. Nitisak N, Jiraphon S (2019) Vibration reduction of an overhung rotor supported by an active magnetic bearing using a decoupling control system. Machines 7(73):1–16. https://doi.org/10.3390/machines7040073

# Numerical Study to Evaluate the Flow Behaviour of the Gases for a Downdraft Gasifier

**Shahfaiz Fayaz and Arpit Thakur**

**Abstract** Environmental deprivation and energy demand are presently the dynamic issues for global sustainable development. Since renewable technologies are neat and clean sources of energy and optimum use of these resources restrict environmental effects, produce the least secondary wastes and are bearable based on current and future economic and social needs. This paper deals with the flow regime of the product gases (carbon dioxide, oxygen and carbon monoxide) which are obtained from paper waste and animal waste through a combustion process using a two-dimensional downdraft gasifier having an internal diameter 0.1 m and height of 1 m were analysed using CFD Software. Both biomasses with a feed rate of 0.1 and 0.3 kg/h with airflow rate were fed into the gasifier. The injection of the biomass discrete phase model (DPM) is used. The product gases after combustion of different biomass can be brought for the combined heat power (CHP) and electricity generation. The proximate and ultimate analysis of biomasses was obtained and calculated to generate stoichiometry and reaction mechanisms using the coal calculator in the species transport code. The results of the study show the optimum conditions to operate a down draft gasifier and effective handling of the product of combustion for limiting the wastage of these gases so that it can be used for other applications; moreover, it will reduce the adverse effect and it was offering to the environment. But no particular attention was paid to char conversion.

**Keywords** Discrete phase model · Coal calculator · Stoichiometry · Species transport

S. Fayaz · A. Thakur (✉)
Department of Mechanical Engineering, Lovely Professional University, Phagwara, Punjab 144401, India
e-mail: arpit.19469@lpu.co.in



R. Kumar et al. (eds.), *Recent Trends in Thermal Engineering*, Lecture Notes in Mechanical Engineering, https://doi.org/10.1007/978-981-16-3132-0_12

# 1 Introduction

The vitality catastrophe and ecological hardship are presently two powerful issues for worldwide reasonable improvement. Renewable resources viewed as one of the well springs of energy and continuous utilization of these resources delimitate the impacts and will create lesser wastes and are reasonably dependent on present and future economic and social requirements. The basic source of all energies is the sun. The basic source of solar energy is heat and light and the environment absorbs and transforms this basic source. The transformations result in renewable energy flow such as wind energy and biomass. The availability of biomass is traditionally in the solid form. This includes harvests residues, animal waste, urban waste, forestry waste, food waste, plants and trees waste, and vegetal waste. Biomass energy is conceivably the world's biggest and most reasonable vitality hotspots for control age in the twenty-first century traditional. The conversion of this biomass requires a thermal-synthetic procedure in which carbon-containing (HC) materials (coal, oil-coke, biomass and so on.) may be modified over to mixture gasoline (syngas) or maker gas by using techniques non-stoichiometric (partial oxidation) with steam, air and oxygen. The tool which plays out this effort is known as the gasifier. And the downdraft gasifier is favoured due to the flexible adaptation of gas production to loading and low sensitivity to char, dust and tar conversion. By choice biomass chosen for the conversion into product gases for combustion and gasification is animal waste and paper waste. The resultant gases carbon dioxide, oxygen and carbon monoxide are obtained after the execution of the combustion process in ANSYS Workbench and after the combustion of different biomass can be brought for the combined heat power (CHP) and electricity generation.

# 2 Methodology

## 2.1 *Gasifier Dimensions*

The design of the downdraft bed was made on a pilot scale for CFD analysis according to the literature with some changes in the inlet and outlet portions. Downdraft gasifiers are the simplest, and most common types of a gasifier for biomass and the dimensions are given in Table 1.

**Table 1** Calculated dimensions for downdraft gasifier

| S no | Parts | Dimensions (mm) |
|---|---|---|
| 1 | Internal diameter | 100 |
| 2 | Height | 1000 |
| 3 | Biomass inlet | 100 |
| 4 | Air inlet | 50 |
| 5 | Outlet | 100 |

## 2.2 *Mass Balance*

The mass balance generally comprises the design for the gas flow rate and feed rate. The mass balance of a thermodynamic system effort to establish the equivalence of input and output material streams. We can also use this parameter for design and calculation if the anticipated presentation of the gasifier can be assessed from initial pilot/laboratory-scale studies. The mass flow rate of feed and air is given in Table 2.

# 3 Design and CFD Analysis

The ANSYS Workbench was approached towards the designed geometry and FLUENT towards the analysis of product gases as shown in Fig. 1.

The objective of this study is to use different biomass to obtain product gases for electricity generation and CHP. Computational fluid dynamics involves complex chemical reactions and multiphase flows for dealing with gasification and combustion processes with a strong connection of turbulence and radiation. Computational fluid dynamics codes would behave towards biomass particles as a discrete phase and for the gas, species would act towards a continuous phase. For modelling biomass reactions in CFD, the typical parameters/codes were evaluated including its devolatilization, char oxidation and gas-phase reactions, turbulence and radiation sub-models. Therefore, by switching these parameters CFD analysis provides convenient data and results.

The ultimate and proximate analysis of animal waste and paper is taken from the literature (LasseRosendahl, 2013 and PrabirBasu, 2010) given in Tables 3 and 4 represents the approach of CFD towards combustion and gasification. And the HCV was obtained from the reaction known as DULONG, S formula, i.e. HCV = (33,800 C + 144,000 ($H_2$ – $O_2$/8) + 9,270 S) KJ/Kg. Supply of inputs for proximate and ultimate analysis generating the gas-phase reaction of paper biomass.

$$C - 0.5H_2 \cdot 29O0.42\,N0.047\,S0.030 + 0.17O_2 \Rightarrow$$
$$0.46CO + 0.27H_2O + 0.0237N_2$$
$$+ 0.03SO_2\,C0 + 0.5O_2 \Rightarrow CO_2$$

**Table 2** Mass flow rate for biomass/air injection

| Biomass | Feed flow rate (kg/s) | Air flow rate (kg/s) | Temp (k) |
|---|---|---|---|
| Animal waste | 0.1 | 0.3 | 500 |
| Paper waste | 0.1 | 0.3 | 500 |

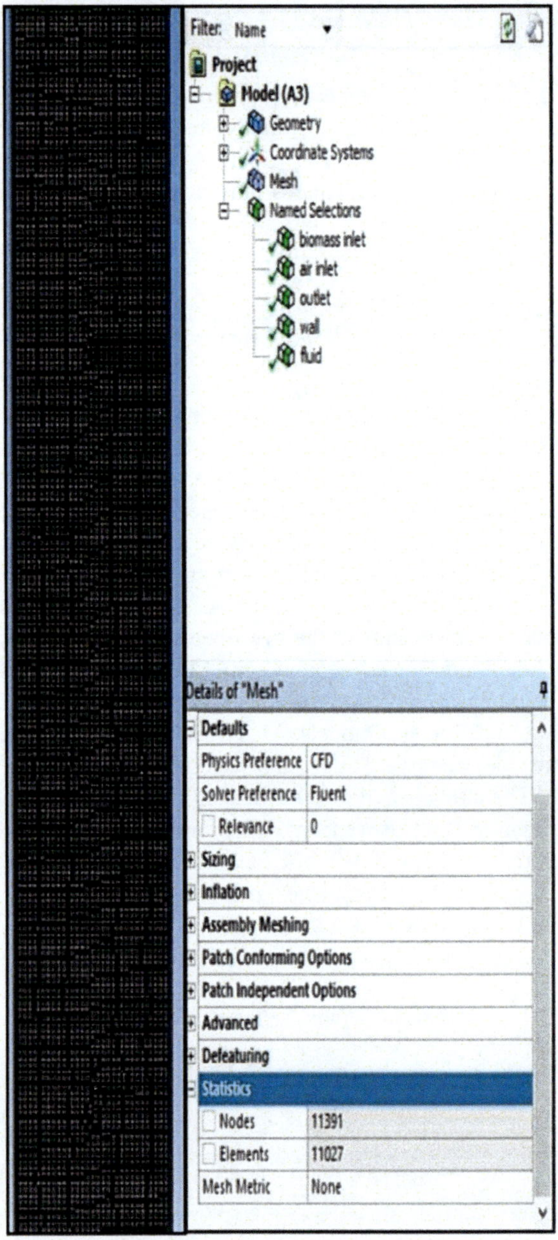


**Fig. 1** Individual reactor design parameters using Workbench and meshed model with assigned boundary names

The total number of two reactions with two reactants and three products in the reaction first and in the second reaction the number of reactants was two, and the number of products was one having different stoichiometry coefficient and rate

**Table 3** Animal waste and paper composition in mass%

| Biomass characteristics | Animal waste | Plant |
|---|---|---|
| Carbon | 42.4 | 43.7 |
| Hydrogen | 5.5 | 5.8 |
| Oxygen | 31.4 | 44.3 |
| Nitrogen | 2.4 | 0.3 |
| Sulphur | 0.3 | 0.2 |
| Fixed carbon | 60 | 60 |
| Volatile matter | 3.8 | 20 |
| Ash | 17.8 | 6.0 |
| Moisture | 19 | 14 |
| HCV (MJ/Kg Daf) | 23.7 | 18.8 |

**Table 4** Main sub-models considered

| Subject | Sub-model |
|---|---|
| Species | Species transport |
| Biomass particle | Discrete phase model |
| Gas phase | Finite rate/eddy dissipation |
| Turbulence | K-epsilon realizable |
| Radiant | Discrete ordinate |
| Particle dispersion | Stochastic tracking |
| Devolatilization | Two competing rates |
| Char reaction | Volumetric |

**Table 5** Injection properties

| Injection type | Surface |
|---|---|
| Release from surfaces | Inject using the normal direction |
| Particle type | Combusting |
| Material | Wood particle |
| Diameter distribution | Uniform |
| Oxidizing species | Oxygen |
| Evaporating species | Water |
| Devolatilizing species | Volumetric |
| Product species | Carbon dioxide |

exponent for each product and reactants. DPM model was used to inject biomass and was released from the surface, and particle type was taken combusting with a diameter of 0.1 m and temperature 500 K as shown in Table 5. Table 6 presents the given boundary conditions.

**Table 6** Defining the boundary conditions

| | |
|---|---|
| Reference frame | Absolute |
| Mass flow specification method | Mass flow rate |
| Mass flow rate (kg/s) | 0.1 (constant) |
| Initial gauge pressure (pascal) | Atmospheric |
| Direction specification method | Normal to boundary |
| Radiation external black body temp. method | Boundary temperature |
| Specification method | Turbulent intensity % 5 and turbulent viscosity ratio 10 |
| DPM | Reflect for airflow and trap for biomass |

# 4 Results

After the completion of the simulation results, i.e. contours of the mole fraction of flow regime of carbon dioxide, oxygen and carbon monoxide, and temperature were displayed. It was observed at the production of product gases produce discomparatively rich in carbon dioxide and oxygen, and a small amount of carbon monoxide from both animal waste and paper biomass. The production of a favourable amount of carbon dioxide and nitrogen is the result of the process of partial combustion which takes place with gasification of biomass simultaneously. The thermal combustion and gasification of biomass include a set of compound chemical reactions that leads to the formation of three fractions: the product gases, ashes/char, and condensates and the most important fraction is involving almost 70% of these lightweight gases. Also, the effect of temperature on biomass gasification and combustion has a most influential effect on them as it has been observed through contours that the increasing temperature in the gasifier leads the improved and the gas yield.

In general, models predict the main gaseous species at the outlet reasonably well, but the validation is supported only for the few species and biomasses as the literature is taken for the kinetics vary greatly from one source to other literature. Therefore, only $CO_2$, CO, $O_2$ are the only species needs to be validated as per the (2017) by Daniel Owusu Asante is concerned he has used different gasifier and its dimensions, and the calculation for mass flow rate of air and fuel is differential though results of product gases, i.e. mole fraction of $CO_2$, CO, $O_2$ is satisfactory. The temperature distribution is exact same as per in his case. In the case of Daniel Owusu Asante temperature also reaches up to 900 K.

Figures 2 and 3 show the mole fraction of carbon dioxide, and it is evident that mole fraction is highest at outlets in both cases and here feed as well as gas travel downwards. And distribution remains almost constant in both cases

Figures 4 and 5 show the mole fraction of oxygen of downdraft gasifier is showing instantaneous consumption from top to bottom in case of animal waste combustion while a little deviation and unstable distribution across the gasifier.

**Fig. 2** Carbon dioxide from animal waste

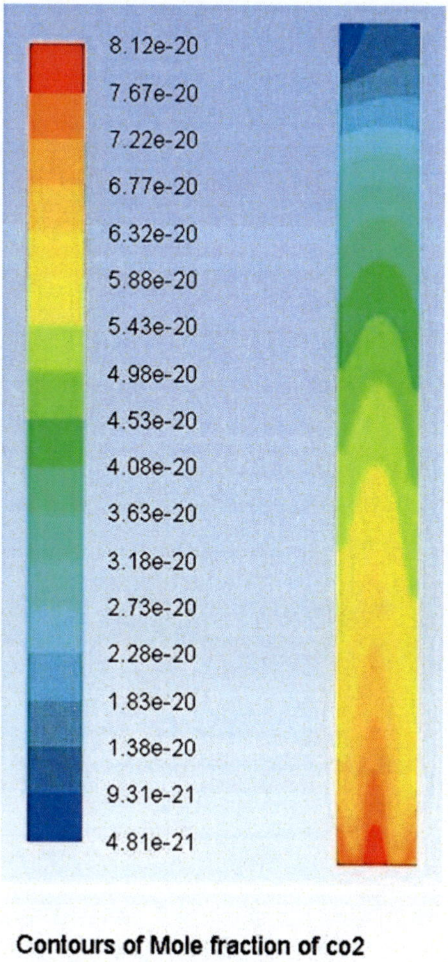


Figures 6 and 7 show the mole fraction of carbon monoxide. The distribution of carbon monoxide in the animal waste and paper biomass gasifier showing throughout most of the gasifier. With the increase in the temperature, higher the production of carbon monoxide.

Figure 8 presents the temperature distribution in the downdraft gasifier in the case of animal waste combustion. The initial temperature at the inlet was given 500 K, and after the execution of the simulation, the temperature is reached up to 800 K at the outlet. The distribution of temperature rises with a fall of height. In all the cases, the temperature distribution throughout the gasifier remains convenient and higher in the combustion zone and rises with a fall in height up to the outlet chamber.

**Fig. 3** Carbon dioxide from paper biomass

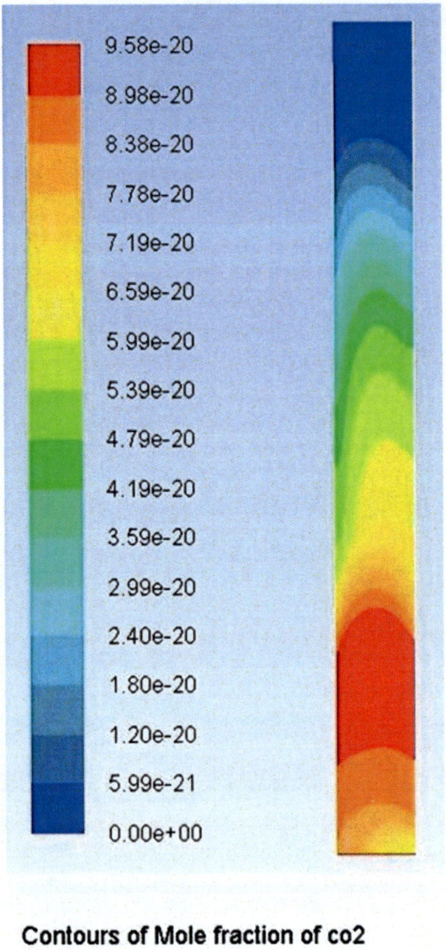


Figure 9 presents the temperature distribution in the downdraft gasifier in the case of plant biomass combustion. The temperature distribution across the gasifier is almost stable and reaches up to 900 K.

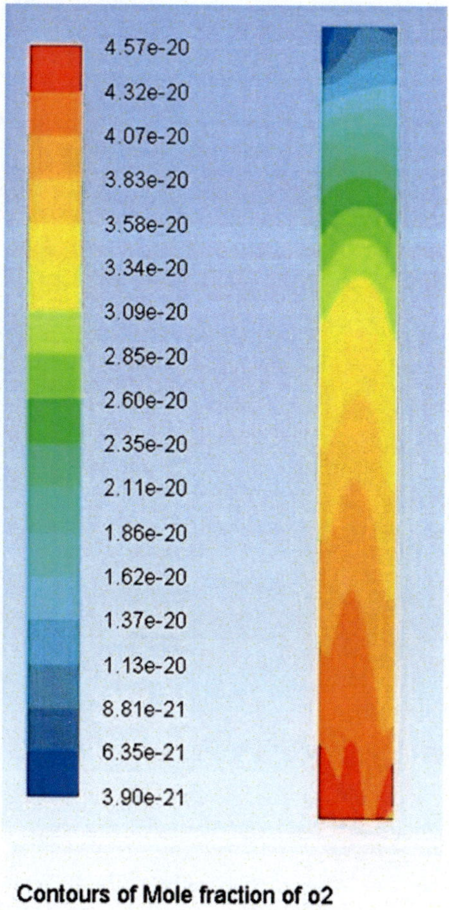


**Fig. 4** Oxygen from animal waste

**Fig. 5** Oxygen from paper biomass

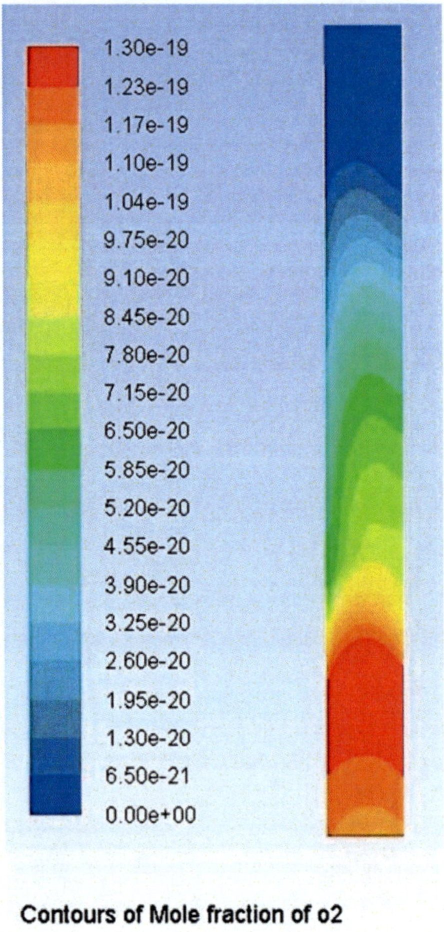

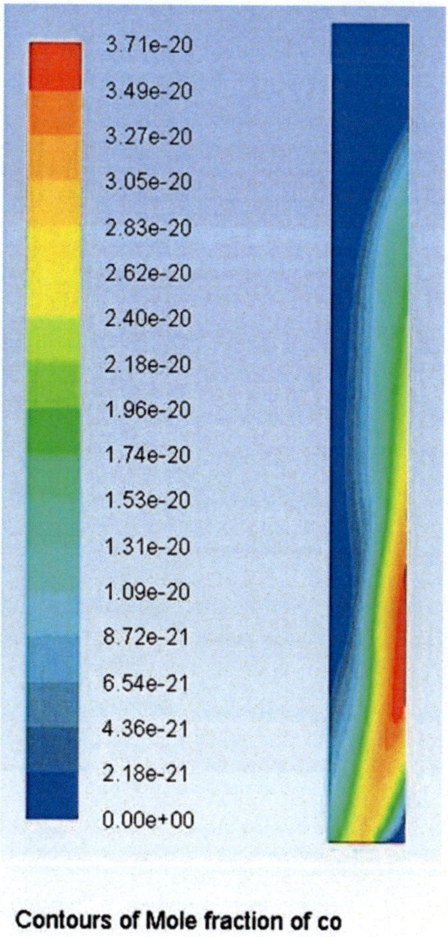


**Fig. 6** Carbon monoxide from animal waste

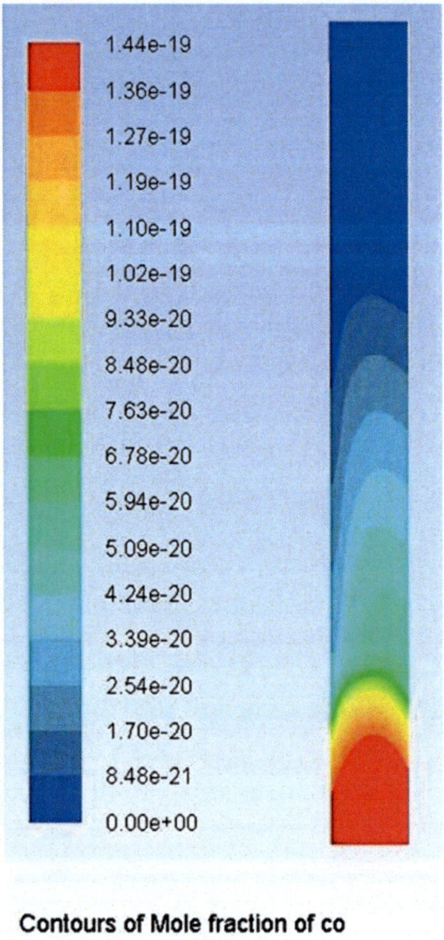


**Fig. 7** Carbon monoxide from paper biomass

**Fig. 8** Temperature from animal waste

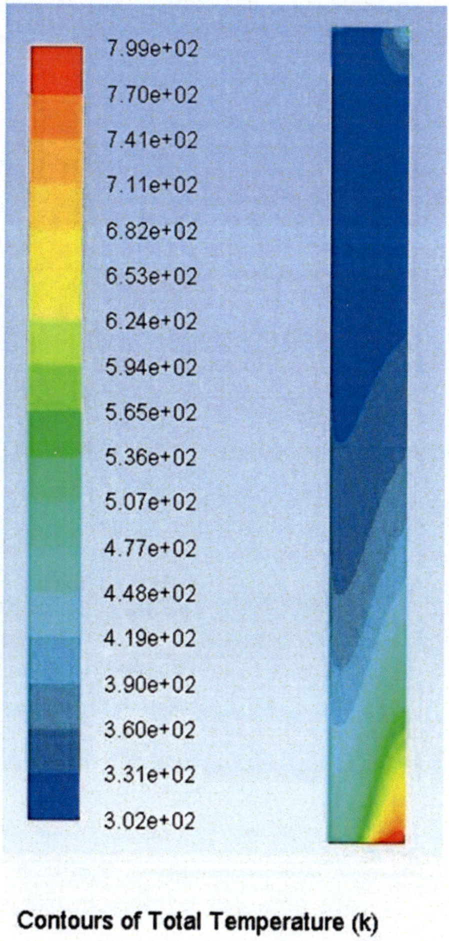

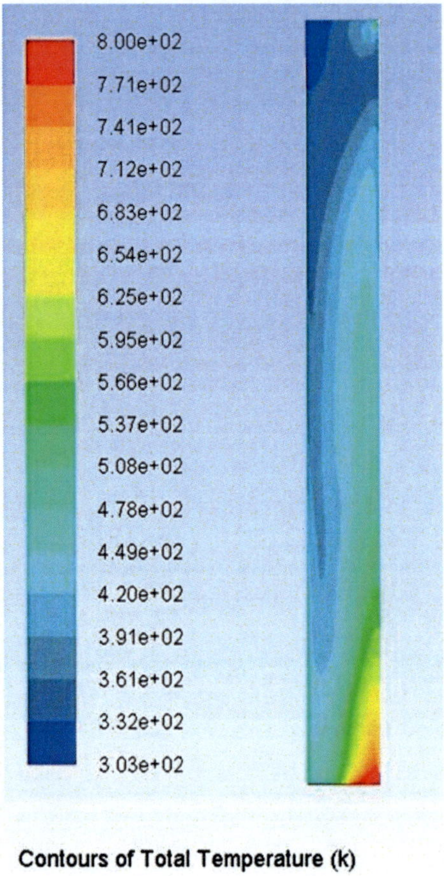


**Fig. 9** Temperature from paper biomass

## 5 Conclusion

CFD model of downdraft biomass gasifier has been developed and was used for the analysis of combustion and gasification for the extraction of gas compositions. After the execution of the simulation, results of different gases carbon dioxide, carbon monoxide and oxygen were obtained and temperature distribution. Also, no attention was paid towards char/ash conversion.

# Experimental Investigation on Hybrid Liquid Desiccant Cooling System for Hot and Humid Climatic Conditions of India

Check for updates

**Sanjeev Kumar, Faizan Ahmad, and Minesh Vohra**

**Abstract** A liquid desiccant dehumidification system (LDDS) combined with vapour compression refrigeration system (VCRS) has been experimentally studied. The structural packing of dehumidifier/regenerator is done by honeycomb PVC fills and provides the sufficient surface area for proper mixing of desiccant solution and process air for optimum moisture exchange. Calcium chloride ($CaCl_2$) is used as a desiccant to perform the operation of LDDS, and R134a is used as a refrigerant in the VCRS to cool the hot dehumidified air. In this study, moisture removal rate (MRR) and dehumidification effectiveness ($\varepsilon_{deh}$) are calculated to estimate the performance of dehumidifier, and coefficient of performance (COP) is calculated to evaluate the performance of the whole system and VCRS without LDDS. These constraints are considered at different air temperatures and velocity. The maximum increase in COP of system and MRR is 8.09% and 26.75%, respectively, as compared with VCRS only, keeping constant air temperature of 35 °C. The maximum dehumidification effectiveness is decreased by 26.32% for the same case.

**Keywords** Liquid desiccant dehumidification system · Dehumidification effectiveness · Moisture removal rate · Coefficient of performance · Liquid desiccant

## 1 Introduction

Energy deficiency is the major problem of the growing world, therefore a necessity to create the novel energy-saving as well as eco-friendly air-conditioning systems. A liquid desiccant cooling system (LDCS) utilizes low-grade energy can satisfactorily achieve those requirements. Desiccants are used to operate LDCS, and it has the property of absorbing and holding moisture from the process air when brought into contact with each other because they have a high attraction towards

S. Kumar (✉) · F. Ahmad · M. Vohra
Lovely Professional University, Phagwara, India
e-mail: Sanjeev.17704@lpu.co.in



R. Kumar et al. (eds.), *Recent Trends in Thermal Engineering*, Lecture Notes in Mechanical Engineering, https://doi.org/10.1007/978-981-16-3132-0_13

water vapour. In the past, Jain et al. [1] concluded that the higher values of wetness factor (*Fw* and *Fh*) provide the more uniform wetting which enhances the performance of the system. Dai et al. [2] suggested that the lower electricity consumption, higher COP of the system and reduced size of VCS, etc. Sultan et al. [3] concluded that the moisture transfer rate from the desiccant was improved by a rise in inlet air flow rate and inlet solution temperature. Fumo and Goswami [4] results show that the mass flow ratio of air to the desiccant solution varied from 0.15 to 0.25 for air dehumidification as well as desiccant regeneration which was lesser than the MR values of 1.3–3.3 used in most other studies. Elsarrag et al. [5] concluded that high liquid flow rates do not affect the performance of the system if the value of the ratio of liquid to airflow ratio does not exceed 2, and reduction in wet-bulb temperature in the dehumidifier was achieved in the range of 4.5–9 °C. Yin et al. [6] suggested that the average mass transfer coefficient of the packing regenerator was 4 g/($m^2$ s) based on experimental results. Jain and Bansal [7] concluded that large deviations in dehumidification effectiveness values that varied from 10 to 50% or more when larger variations happening for lesser ratios of liquid to gas flow rates. Liu et al. [8] results show that the water vapour elimination rate rises with growing air flow rate, desiccant flow rate and inlet desiccant temperature, reduces with humidity ratio of inlet air and inlet desiccant concentration and changes minute with inlet air temperature. Mohan et al. [9] concluded that high dehumidification in the absorber was achieved at high humidity ratio and low temperature of the inlet air. Zhang et al. [10] concluded that with the increase in air velocity from 0.5 to 1.5 m/s, mass transfer coefficient in the dehumidifier and regenerator was varied from 4.0 to 8.5 g/$m^2$s and 2.0 to 4.5 g/$m^2$s, respectively. Zhang et al. [11] suggested that the COP of the hybrid system was enhanced by 20% and 100% in summer and winter, respectively, as compared to a traditional air-conditioning system. Bansal et al. [12] concluded that the maximum effectiveness and moisture removal rate were achieved within internal cooling. Bassuoni [13] suggested that the increase of mass transfer coefficient and MRR for both dehumidifier/regenerator was established by increasing both air and solution flow rates. Yin et al. [14] concluded that the thermal efficiency of regeneration would be optimum when the ratio of air to desiccant flow rate was around 8. Bergeron and Chiari [15] simulations' results show that significant energy savings of more than 50% at high latent load in the conditioned environment. Many researchers had experimentally examined the liquid desiccant cooling system with dissimilar kinds of structural packing in a dehumidifier. But no research work has been done on the packing of the dehumidifier with PVC fills and LDDS combined with VCRs. The performance of an LDDS combined with VCRS is presented. The objective of this manuscript is to produce dry and cooled air according to the required indoor conditions. The performance parameter such as moisture removal rate (MRR) has been evaluated experimentally.

## 2 Experimental Set-Up

Figure 1 illustrates the photographic view of the experimental set-up. It comprises of two sub-systems, namely liquid desiccant dehumidification system (LDDS) and vapour compression refrigeration system (VCRS). LDDS is used to dehumidify the process air while VCRS is used to decrease the temperature of dehumidified air for better cooling. The key components used in the experimental setup are dehumidifier, regenerator, liquid desiccant, desiccant solution heating and cooling chamber, air heater, solution pump, axial fan, dehumidification chamber, connection pipes and VCRS.

## 3 System Operation

The schematic diagram of the anticipated system is shown in Fig. 2. In dehumidifier, the liquid desiccant absorbs the water vapour present in the process air, and after dehumidification process, the dehumidified air at the outlet of dehumidifier is passed to the evaporator where the heat of air taken away by the refrigerant (R134a) to cool the air. Liquid desiccant gets diluted and collected in the dehumidifier tank after absorbing the water vapour from the process air. To get the original state of the diluted liquid desiccant solution, it is regenerated in the regenerator. The diluted liquid desiccant solution is pumped to the solution heating chamber, where its temperature is raised around 50 °C, and afterwards, the hot desiccant solution is sent to a regenerator in which its losses the water vapour to the surrounding air passing through it and regenerated solution in concentrated form is collected into

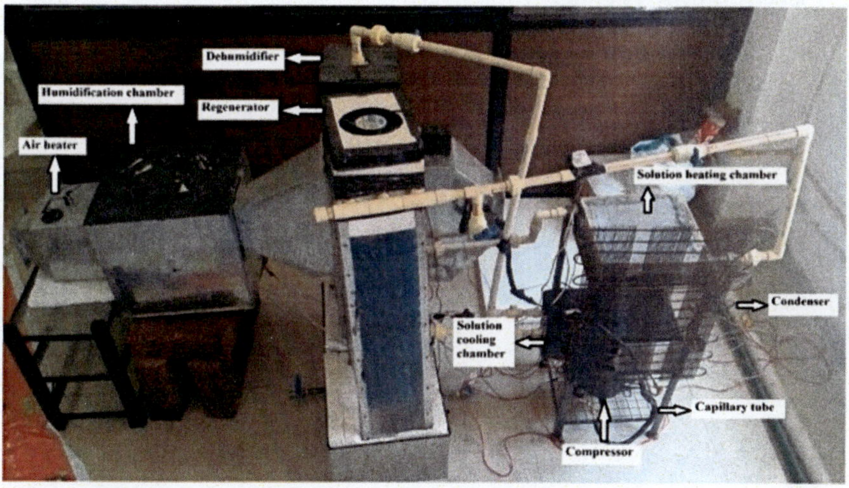


**Fig. 1** Photographic view of the experimental set-up

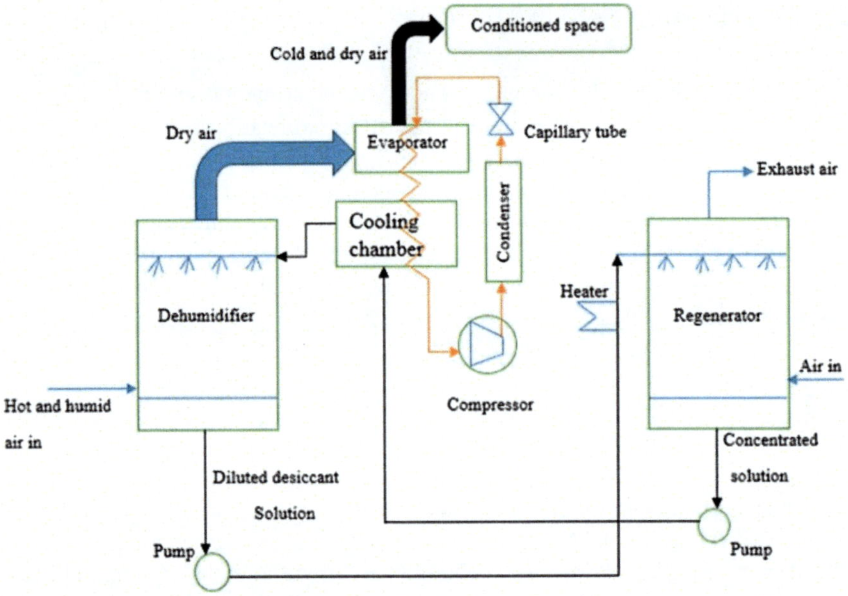


**Fig. 2** Schematic diagram of the proposed system

the regenerator tank. The concentrated solution is pumped to the desiccant solution cooling chamber where it gets cooled by refrigerant coming from the evaporator coil, and after that, it is a pass to the dehumidifier, and thus, the cycle continues.

## 4 Results and Discussion

The system performances are examined by calculating parameters like moisture removal rate (MRR), dehumidification effectiveness ($\varepsilon_{deh}$), COP of system and COP of VCRS without LDDS. The experiments are conducted under three cases to estimate the performance of the system.

Case 1: Saturated air with a different temperature at a constant air velocity.

Case 2: Saturated air with different velocities at constant air temperature.

Case 3: Comparison of COP of system and COP of VCRS without LDDS under different conditions.

## *4.1 Saturated Air with a Different Temperature at a Constant Air Velocity*

In this case, the saturated air temperature varies from 30 to 40 °C with the help of an air heater by keeping the constant air velocity of 4.0 m/s. The performance parameters like moisture removal rate (MRR), dehumidification effectiveness, COP of system and COP of VCRS without LDDS are calculated at various air temperature.

### 4.1.1 Variation of Moisture Removal Rate (MRR) Dehumidification Effectiveness (DE) at Different Air Temperature

See Fig. 3.

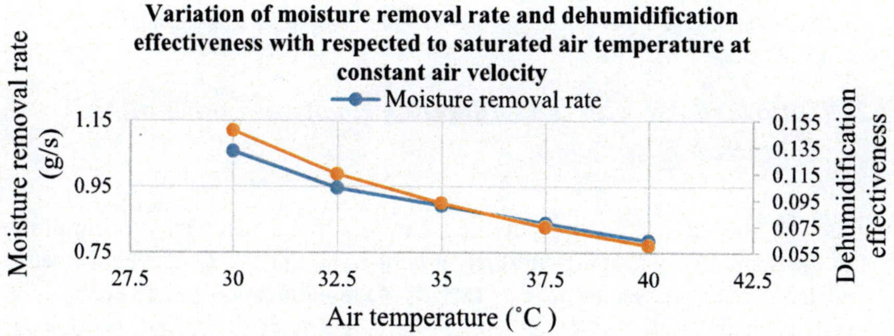


**Fig. 3** Variation of moisture removal rate and dehumidification effectiveness with saturated air temperature

### 4.1.2 Variation of COP of the System at Different Air Temperature

See Fig. 4.

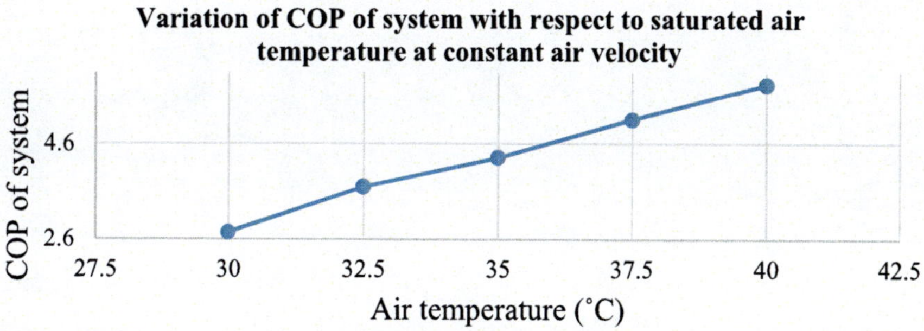


**Fig. 4** Variation of COP of the system with saturated air temperature

#### 4.1.3 Variation of COP of VCRS Without LDDS at Different Air Temperature

See Fig. 5.

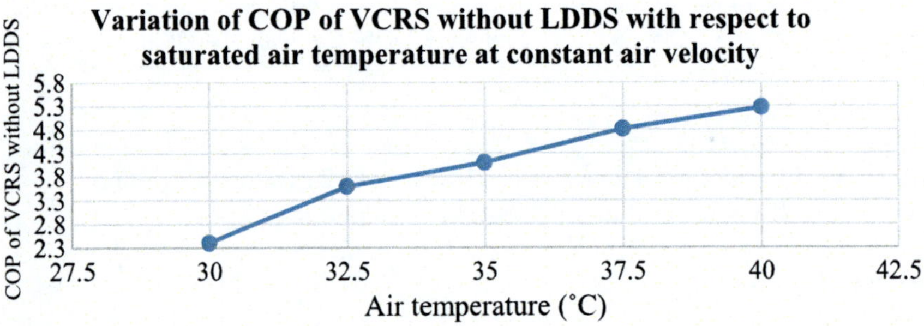


**Fig. 5** Variation of COP of VCRS without LDDS with saturated air temperature

### *4.2 Saturated Air with the Different Velocity at Constant Air Temperature*

In this case, the saturated air velocity varies from 3 to 4.5 m/s with the help of the fan of the regulator by keeping constant air temperature of 35 °C. The performance parameters like moisture removal rate (MRR), dehumidification effectiveness, COP of system, and COP of VCRS without LDDS are calculated at various air temperature.

#### 4.2.1 Variation of Moisture Removal Rate (MRR) and Dehumidification Effectiveness (DE) at a Different Air Velocity

See Fig. 6.

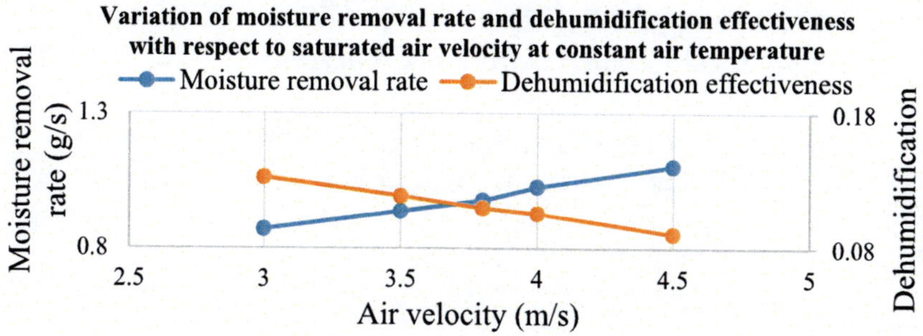


**Fig. 6** Variation of moisture removal rate and dehumidification effectiveness with saturated air velocity

### 4.2.2 Variation of COP of the System at a Different Air Velocity

See Fig. 7.

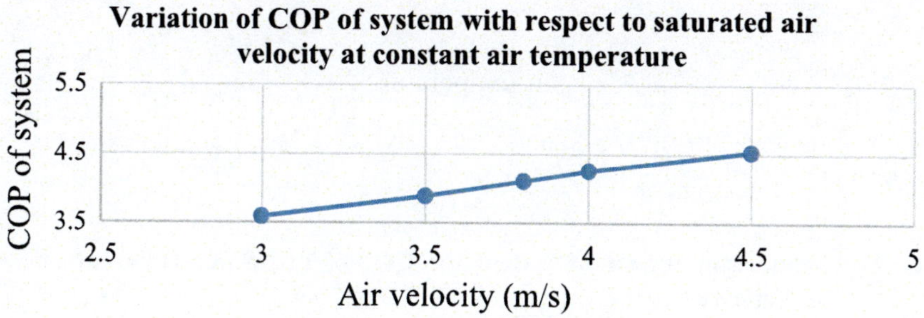


**Fig. 7** Variation of COP of the system with saturated air velocity

### 4.2.3 Variation of COP of VCRS at a Different Air Velocity

See Fig. 8.

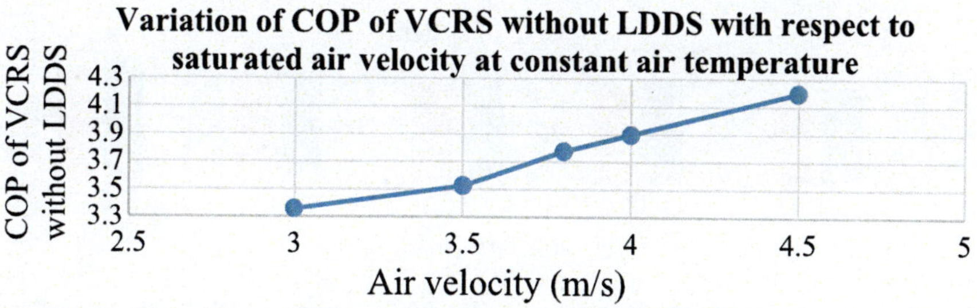


**Fig. 8** Variation of COP of VCRS without LDDS with saturated air velocity

## *4.3 Comparison of COP of System and COP of VCRS Without LDDS Under Different Conditions*

### 4.3.1 Comparison of COP of System and COP of VCRS Without LDDS at Different Air Temperature

See Fig. 9.

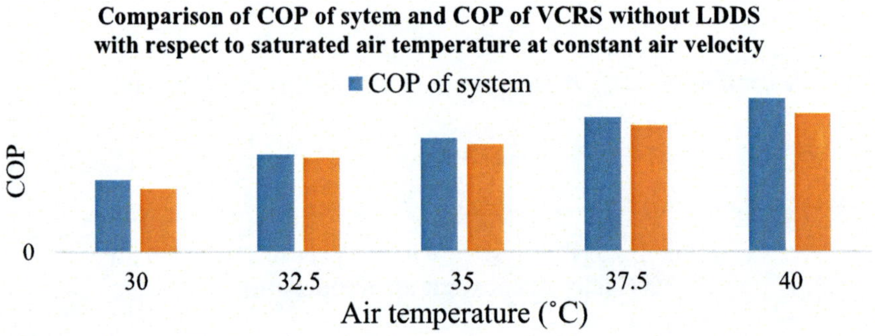


**Fig. 9** Comparison of COP of system and COP of VCRS without LDDS with saturated air temperature

### 4.3.2 Comparison of COP of System and COP of VCRS at a Different Air Velocity

See Fig. 10.

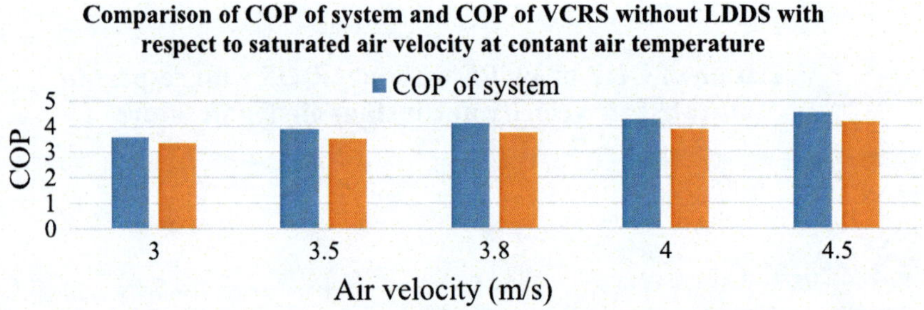


**Fig. 10** Comparison of COP of system and COP of VCRS without LDDS with saturated air velocity

## 5 Conclusions

In the present study, the performance of a liquid desiccant dehumidification system combined with the vapour compression refrigeration system has been analysed. The following conclusions have been drawn from the results:

1. MRR decreases from 1.058 to 0.788 g/s, dehumidification effectiveness decreases from 0.148 to 0.061, COP of system increases from 2.75 to 5.84, and COP of VCRS without LDDS increases from 2.41 to 5.27, when the saturated air temperature varies from 30 to 40 °C at a constant air velocity of 4.0 m/s.
2. MRR increases from 0.871 to 1.104 g/s, dehumidification effectiveness decreases from 0.133 to 0.098, COP of system increases from 3.59 to 4.54, and COP of VCRS without LDDS increases from 3.36 to 4.20, when the saturated air velocity varies from 3 to 4.5 m/s by keeping constant air temperature of 35 °C.
3. The maximum increase in COP of system and MRR is 8.09% and 26.75% as compared with VCRS only, keeping constant air temperature of 35 °C. The maximum dehumidification effectiveness is decreased by 26.32% for the same case.
4. The enhancement in the design of components of hybrid cooling systems for better performance and nominal losses is important to consider. Supplementary changes that can be employed in the proposed experimental set-up to improve the performance of the system are:
    - The internal cooling arrangement can be employed in the dehumidifier to counter the heat generation during moisture absorption and better dehumidification.
    - Drift eliminator can be used to reduce the problems of carry-over of desiccant in case of structural packed bed dehumidifier/regenerator.

## References


1. Jain S, Dhar PL, Kaushik SC (2000) Experimental studies on deh umidifier and regenerator of a liquid desiccant cooling system. Appl Therm Eng 20:253–267
2. Dai YJ, Wang RZ, Zhang HF et al (2001) Use of liquid desiccant cooling to improve the performance of vapour compression air conditioning. Appl Therm Eng 21:1185–1202
3. Sultan GI, Hamed AM, Sultan AA (2002) The effect of inlet parameters on the performance of packed tower-regenerator. Renew Energy 26:271–283
4. Fumo N, Goswami DY (2002) Study of an aqueous lithium chloride desiccant system: air dehumidification and desiccant regeneration. Solar Energy 72:351–361
5. Elsarrag E, Elmagzoub E, Ali M et al (2005) Design guidelines and performance study on a structured packed liquid desiccant air-conditioning system. HVAC&R Res 11:319–337
6. Yin Y, Zhang X, Chen Z (2007) Experimental study on dehumidifier and regenerator of liquid desiccant cooling air conditioning system. Build Environ 42:2505–2511
7. Jain S, Bansal PK (2007) Performance analysis of liquid desiccant dehumidification systems. Int J Refrig 30:861–872

8. Liu XH, Jiang Y, Chang XM, Yi XQ (2007) Experimental investigation of the heat and mass transfer between air and liquid desiccant in a cross-flow regenerator. Renew Energy 32:1623–1636
9. Shaji Mohan B, Prakash Maiya M, Tiwari S (2008) Performance characterization of liquid desiccant columns for a hybrid air-conditioner. Appl Thermal Eng 28:1342–1355
10. Zhang L, Hihara E, Matsuoka F et al (2010) Experimental analysis of mass transfer in adiabatic structured packing dehumidifier/regenerator with liquid desiccant. Int J Heat Mass Transfer 53:2856–2863
11. Zhang L, Dang C, Hihara E (2010) Performance analysis of a no-frost hybrid air conditioning system with integrated liquid desiccant dehumidification. Int J Refrig 33:116–124
12. Bansal P, Jain S, Moon C (2011) Performance comparison of an adiabatic and an internally cooled structured packed-bed dehumidifier. Appl Therm Eng 31:14–19
13. Bassuoni MM (2011) An experimental study of structured packing dehumidifier/regenerator operating with liquid desiccant. Energy 36:2628–2638
14. Yin Y, Li S, Zhang X et al (2011) Feasibility and performance analysis of a desiccant solution regenerator using hot air. Energy Build 43:1097–1104
15. Bergero S, Chiari A (2011) On the performances of a hybrid air-conditioning system in different climatic conditions. Energy 36:5261–5273

# Power Generation from Wind Using Bladeless Turbine

**Ajay Kumar Kaviti and Amit Kumar Thakur**

**Abstract** Among other renewable energy sources, harnessing wind energy is the least expensive method. A fundamentally different approach to capture wind energy by further reducing prices is being used by Bladeless Wind Power production. The device absorbs the energy of the vortex, an aerodynamic influence. If the wind flows through a buff body, it changes its flow and creates a cyclical vortex pattern. The fixed structure begins to oscillate until the forces caused by these vortices are strong enough. The model maximizes the resulting oscillation and captures the energy instead of preventing these aerodynamic instabilities. Instead of the usual tower, nacelle, and blades as used in conventional turbines, the device has a fixed mast to capture wind energy, a power generator, and a shaft. The purpose of this paper is to ameliorate the understanding of this technology by developing a simulation model, considering parameters like wind velocity.

**Keywords** Vorticity · Frequency · Velocity · Bladeless turbine

## 1 Introduction

In this moment of global market volatility, in order to sustain economic and social development and to create a better standard of living, especially in developing countries, the world desperately needs energy in growing quantities. But the supply of this energy across the globe comes with obligation and dedication. Fulfilling the growing energy needs in a safe and environmentally responsible manner is a great issue. Here is where renewable energy comes in, since it is unquenchable, lessens

A. K. Kaviti
VNR Vignana Jyothi Institute of Engineering and Technology, Hyderabad, Telangana, India

A. K. Thakur (✉)
School of Mechanical Engineering, Lovely Professional University, Phagwara, Punjab, India
e-mail: amit.25010@lpu.co.in

 
R. Kumar et al. (eds.), *Recent Trends in Thermal Engineering*, Lecture Notes in Mechanical Engineering, https://doi.org/10.1007/978-981-16-3132-0_14

conventional energy reliance, is sustainable, and delivers energy independence [1]. In India, the field of renewable energy is an established one, supplying up to 18.37% of the installed power capacity in 2017.

In both indoor and outdoor settings, wind energy is a possible power source that is pervasive. It is obtained by converting the kinetic energy of moving wind into electrical power [2]. It is more prevalent since it is cost-effective, reliable, predictable land-based, and one of the lowest-priced energy sources available today. As of 28-02-2016 wind energy contributes a staggering amount of 56.8% energy in the renewable power supply of the Indian economy [3].

Harnessing energy from wind has various methods; the conventional one is to connect the center shaft of the windmill to the generator. This conversion has losses in energy, and to avoid this, we can switch to bladeless windmills, which utilizes vortex-induced vibrations (VIV) [4]. The bladeless windmill has less moving parts than the conventional windmill, hence requires less maintenance than the conventional windmill. It requires less area and wind speed for its area [5].

Adding to this, limit period oscillations may develop due to the interaction of fluid–structure whenever a particular structure is exposed to wind flow. Utilizing different conversion methods such as electrostatic, electromagnetic, and piezoelectric transitions, the vibration strain energy may be productively converted to electrical energy.

The focus of this article is to provide researchers from different disciplines involved in small-scale wind energy harvesting with valuable feedback and to help them develop a superior comprehension of this methodology. By integrating piezoelectric components into the connecting rod, the energy produced by the vibrations is transformed into electrical energy.

## 2 Bladeless Turbine

### *2.1 Modeling Using CATIA V5*

A three-dimensional model as shown in Fig. 1 is prepared using CATIA V5 software. It consists of a mast made of lightweight materials (PVC, FRP, or other plastics) to capture the wind energy.

The rod is made of flexible materials that oscillate due to the forces acting on the mast. Therefore, energy can be harvested from the rod. The simplicity of the structure and absence of moving parts like bearings and gears reduces manufacturing, transport, and maintenance costs drastically [6].

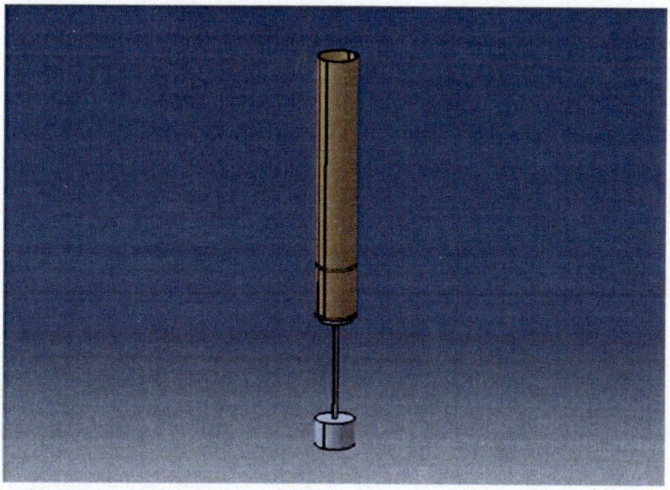

**Fig. 1** 3D model of bladeless wind turbine

## 2.2 *Analysis Using ANSYS Fluent*

The analysis is carried out in a fluent workbench in ANSYS 18.2 software. Considering the symmetric nature of the setup, the computational domain is simplified to a 2D geometry. It is 1,200 mm long and 600 mm wide and the cylindrical mast is placed at a distance of 300 mm from one end. The rectangular side closer to the cylinder is the inlet and away from the cylinder is the outlet. The other two rectangular sides are taken as walls. The geometry is created using the design-modeler workbench in ANSYS software and then the meshing of the geometry is carried out. A triangular dominant mesh is created. The element face sizing around the cylinder wall is set to 1 mm, and the face size is gradually increased up to 50 layers with a growth rate of 1.5 using the inflation command. The global mesh size has a maximum face size of 10 mm. A total of 14,643 nodes, 20,919 elements, and 35,045 2D interior faces are generated for the above specifications. The geometry and meshed model are shown in Fig. 2a, b, respectively. The fluent solution is then calculated with double precision and a single processor.

A transient analysis is carried out because von Karman street or flow-induced vortices is a transient effect. A realizable K-epsilon model is considered. The material of the cylinder is selected as PVC with a density of 1.38e−6 kg/m$^3$ and a surface roughness of 0.015 mm. The fluid is taken as air with a density of 1.225 kg/m$^3$ and viscosity of 1.7894e−05 kg/m-s. The shear condition at the fluid–cylinder boundary is taken as no-slip while at the fluid–wall boundary is taken as zero shear stress to reduce any effects due to fluid interaction with the wall. The inlet velocities are varied between 0.5 and 4 m/s to obtain the frequency of vortices.

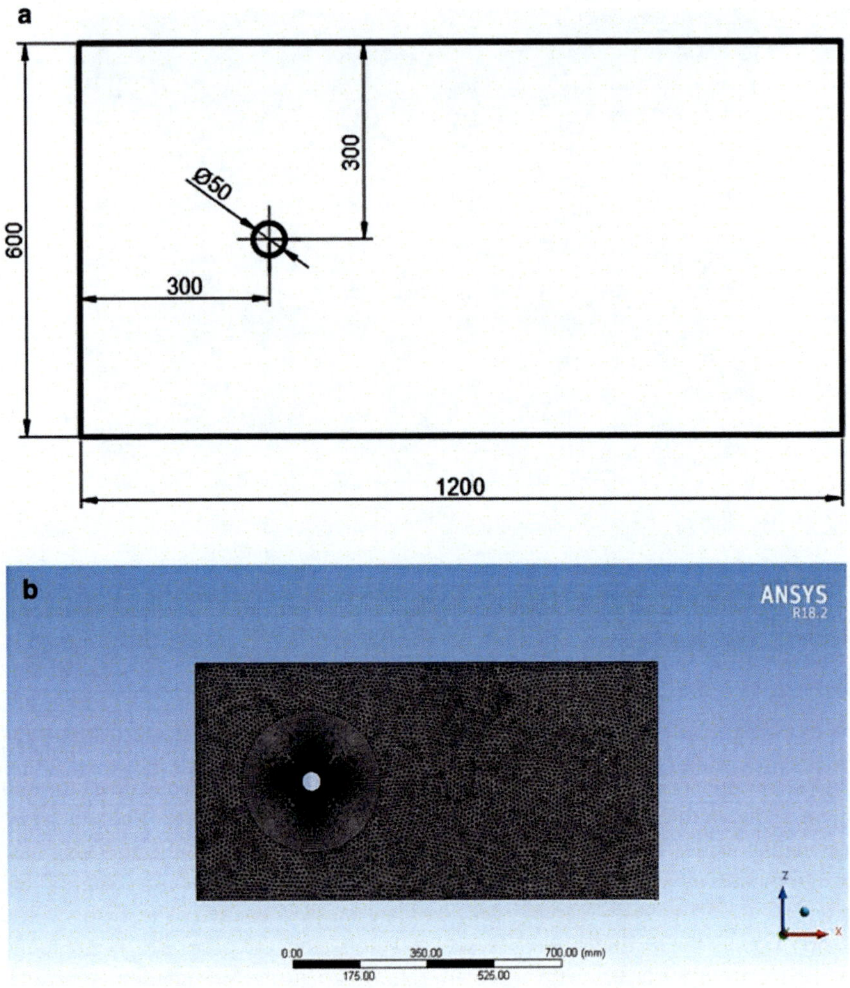


**Fig. 2** **a** Geometry of computational domain (all dimensions are in mm). **b** Meshed domain used for analysis

A second-order implicit formulation is used for transient formulation. A lift-coefficient and drag coefficient plots and files are generated. The calculation is performed for 500 time steps with a time step size of 0.02 s and a maximum of 25 iterations per time step to obtain an accurate result.

## 3 Results and Discussion

Figure 3 represents the formation of vortices for different velocities after a flow time of 10 s or at the end of the simulation. It can be observed that vortices are generated for wind velocities of 0.5, 1.0, 1.5, 2.0, and 2.5 m/s but for a wind velocity of 4.0 m/s, vortex shedding phenomenon is absent. This is because the Reynolds number is very large (Re = 13,694.80) which increases the turbulence in the flow and prevents the formation of Karman vortex street.

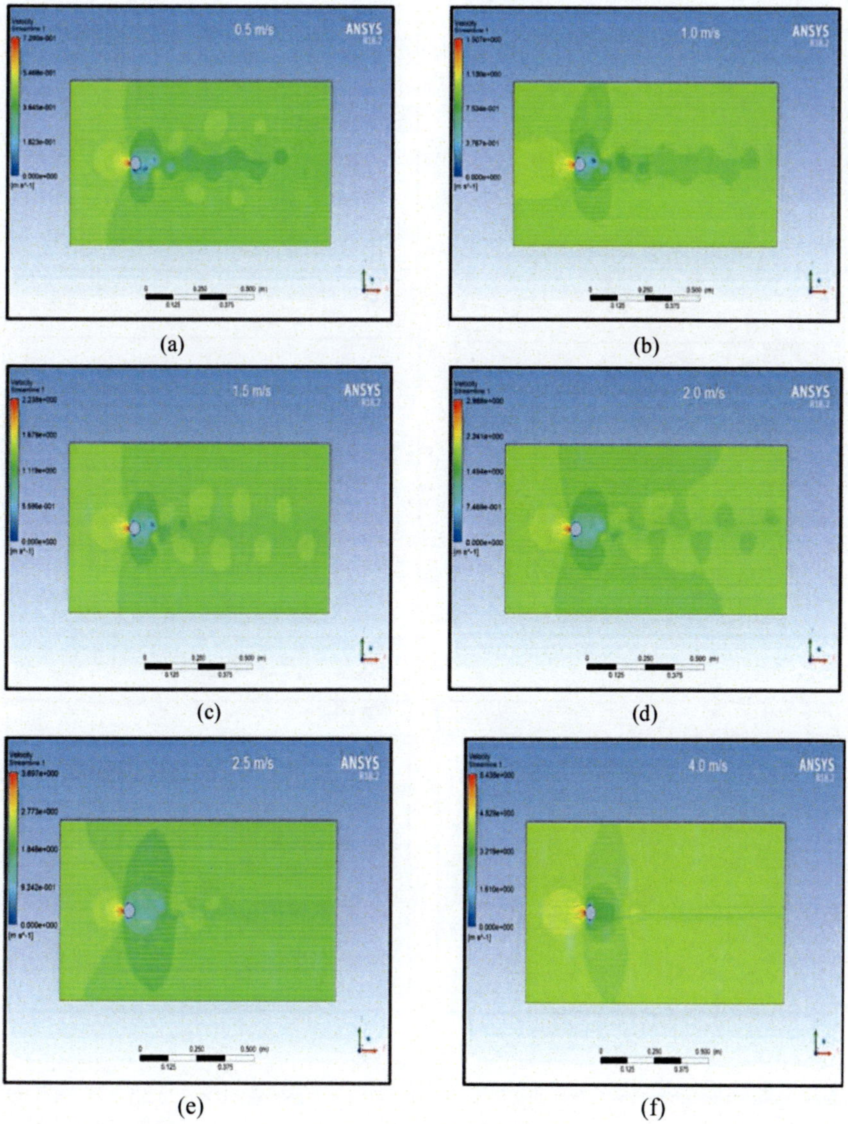


**Fig. 3** Von Karman street **a** 0.5 m/s, **b** 1.0 m/s, **c** 1.5 m/s, **d** 2.0 m/s, **e** 2.5 m/s, **f** 4.0 m/s

Figure 4 represents the flow time(s) against lift-coefficient. The time taken for the formation of stable Karman vortex street can be inferred from these figures. Vortex shedding phenomenon is observed at the end of 3.25 s for a wind velocity of 2.0 m/s whereas for other velocities it takes more than 4.0 s.

The frequencies obtained at 0.5, 1.0, and 1.5 m/s are low. At a velocity of 2.0 m/s, the theoretical frequency obtained is 7.896 Hz (Table 1) while the frequency obtained from ANSYS analysis is 8.5 Hz (Fig. 4d). The percentage error is 7.65% which is

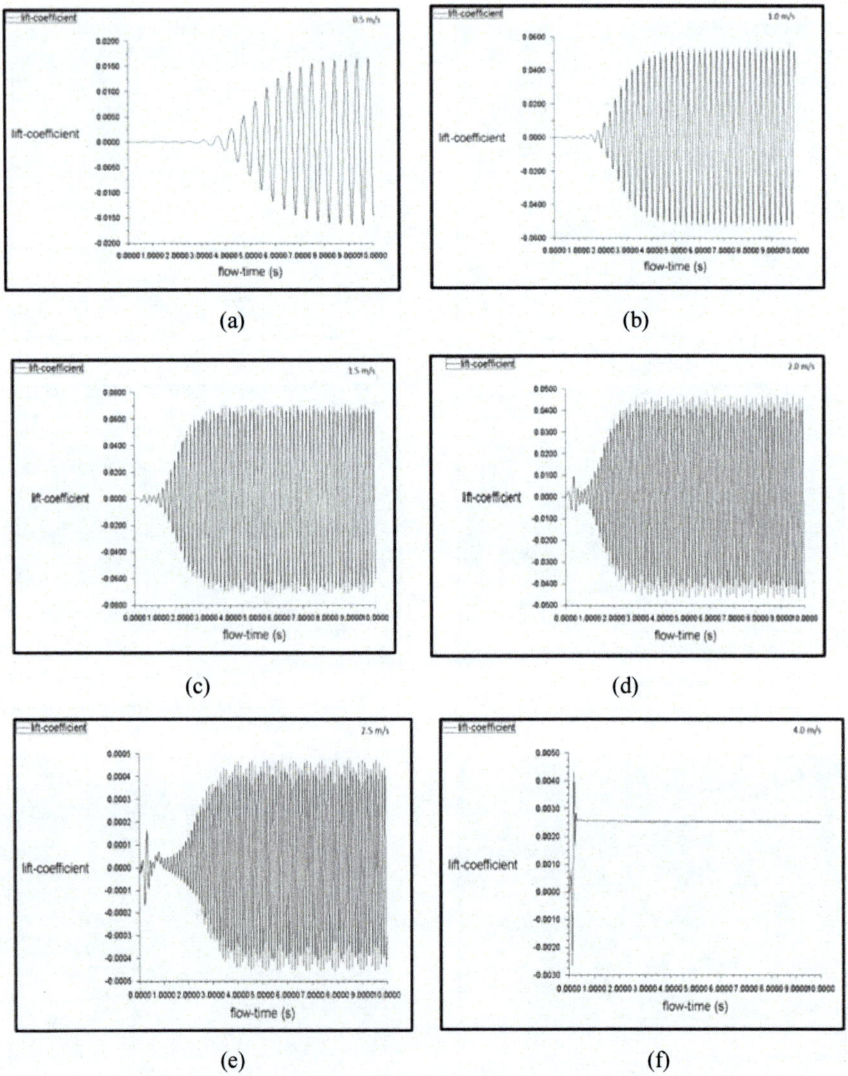


**Fig. 4** Vortex induction with time **a** 0.5 m/s, **b** 1.0 m/s, **c** 1.5 m/s, **d** 2.0 m/s, **e** 2.5 m/s, **f** 4.0 m/s

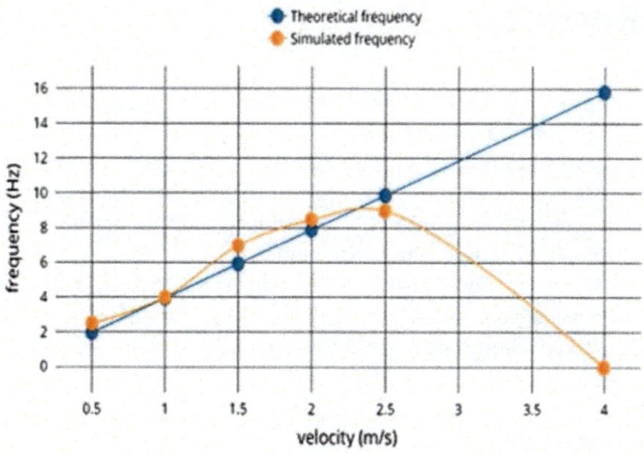


**Fig. 5** Comparison of theoretical and obtained frequencies

**Table 1** Frequencies for various velocities

| S. no | Velocity (m/s) | Reynolds number (Re) | Strouhal number (St) | Theoretical frequency (Hz) | Obtained frequency (Hz) | Error (%) |
|---|---|---|---|---|---|---|
| 1 | 0.5 | 1,711.85 | 0.1957 | 1.957 | 2.5 | 27.75 |
| 2 | 1.0 | 3,423.70 | 0.1968 | 3.936 | 4 | 1.63 |
| 3 | 1.5 | 5,135.55 | 0.1972 | 5.916 | 7 | 18.3 |
| 4 | 2.0 | 6,847.40 | 0.1974 | 7.896 | 8.5 | 7.65 |
| 5 | 2.5 | 8,599.25 | 0.1975 | 9.875 | 9 | 8.86 |
| 6 | 4.0 | 13,694.80 | 0.1977 | 15.816 | 0 | - |

relatively small. At 2.5 m/s, a higher frequency of 9 Hz is observed but the lift-coefficient is very low (around 0.0004).

The average theoretical frequency is 7.566 Hz, and the average obtained frequency is 5.166 Hz. The percentage error is 31.72%. The percentage error can be minimized by further refining the mesh and increasing the nodes and elements.

## 4 Conclusion

ANSYS 18.2 fluid workbench has been adapted to simulate the models.

- The appropriate velocity is found out to be 2.0 m/s because a stable Karman vortex street is observed as soon as 3.25 s after the start of the flow.
- Also, a simulated frequency of 8.5 Hz is obtained which can be used to produce electrical energy.
- The minimum and maximum percentage errors are 1.63% and 27.75% observed at 1.0 m/s and 0.5 m/s, respectively. At the velocity of 2.0 m/s, it is found to be 7.65%. This error can be further reduced by increasing the elements and reducing the time step size.

## References

1. Alrikabi NKMA (2014) Renewable energy types. J Clean Energy Technol 2(1):61–64
2. Shukla V, Kaviti AK (2017) Performance evaluation of profile modifications on straight-bladed vertical axis wind turbine by energy and Spalart Allmaras models. Energy 126:766–795
3. Mann J, Sorensen JN, Morthorst PE (2008) Wind energy. Environ Res Lett 3(1):015001
4. Chaudhari CC, Shriram MA, Unhale SG, Nirmal RS (2017) Fabrication of vortex bladeless windmill power generation model. Int J Sci Technol Eng 3(12):52–56
5. Villarreal DJY (2018) U.S. patent no. 9,856,854. U.S. Patent and Trademark Office, Washington, DC
6. PRACE homepage, www.prace-ri.eu

# Analysis of Shock Shape and Drag Coefficient for Various Spiked Bodies

**R. Rajesh, L. Hethav, M. Devathman, A. Hiren Raaj, and M. Kishore**

**Abstract** This paper mainly focuses on the drag reduction technique of the usage of a forward spike in front of the nose of the body. It has been reported by many researchers the reduction in drag, by the introduction of an aerospike at the point of stagnation. A brief investigation was carried out by varying the Mach number of the free flow fluid and investigating change in the drag with respect to change in Mach number. Various spike configurations have been studied, and the respective shock shapes have been plotted. Mach number has been varied from 4.75 to 6.75 in steps of 0.5, and $K$-$\omega$ (SST) turbulence model with implicit formulation and second-order flux limiter minimum model is used. The computed results reveal a decrease in the drag, with a corresponding increase in Mach number, for all the spike configurations. It is also observed that of the four investigated configurations, the flat aerodisk is most preferred followed by hemispherical aerodisk, cut spike, and sharp spike, respectively, to reduce the drag at a given Mach number.

**Keywords** Spike · Shape · Drag · Density · Recirculation · Steady · Conical · Bow · Aerodisk · Axissymmetric · Hypersonic

## 1 Introduction

High-speed vehicles have to be designed such that there is minimal drag and minimum aerodynamic heating in front of the blunt nose of the body. Various techniques have been found out to change the flow field in front of nose of the body, some of them being the use of a retractable aerospike ahead of the blunt body, and a forward-facing jet at the stagnation point of the nose. An aerospike is a drag reducing device, that creates a detached shock in front of the body. Taking care of the material and thermal stability of the spike alone, instead of the entire blunt head,

R. Rajesh (✉) · L. Hethav · M. Devathman · A. Hiren Raaj · M. Kishore
Department of Mechanical Engineering, Amrita School of Engineering, Amrita Vishwa Vidyapeetham, Bengaluru, India
e-mail: r_rajesh@blr.amrita.edu



R. Kumar et al. (eds.), *Recent Trends in Thermal Engineering*, Lecture Notes in Mechanical Engineering, https://doi.org/10.1007/978-981-16-3132-0_15

serves as a huge advantage. Computational techniques can give us a reasonably agreeable solution. Many researchers across the world have also devoted sufficient time in comparing the experimental solutions with the computed solutions, with them agreeing satisfactorily.

An oblique shock, unlike a normal shock, is inclined at a certain angle with respect to the incident upstream flow direction. A oblique shock occurs when the flow encounters a corner that turns the flow into itself and compresses. A bow shock, also known as a detached shock, is a shock wave that is characterized by its curved propagating disturbance and occurs only when the body is encountered with a supersonic flow, around which the necessary angle of deviation of the flow is higher than the maximum achievable angle of deviation for the attached oblique shock.

## 2 Literature Review

Barzegar Gerdroodbary et al. [1] have analyzed numerically, the effectiveness of using aerodisk/aerospike for blunted cones subjected to a Mach number of 5.75. Results show that hemispherical aerodisk provides the highest drag reduction up to 72% for L/D = 1.5. Miliçev et al. [2] have studied experimentally the influence of Spike Shape at Supersonic Flow Past Blunt-Nosed Bodies and it is concluded that, greater the radial distance of the point of joining (reattachment point) from axis, better are the characteristics aerodynamically. Mehta et al. [3] have numerically investigated the influence of various shapes of spikes mounted on a hemispherical nose of a blunt body and observed that aerodisk exhibits a complex flow field due to its geometry. A study in the use of retractable aerospikes for highly blunted cones was done by Gopalan et al. [4]. The results indicate 40–55% drag reduction for flat and hemispherical aerodisks attached to blunt nose at lower angles of attack.A study done by Sreekanth et al. [5] have investigated the effect of introducing a secondary spike along with a primary aerospike. It is concluded that secondary spike can be used to reduce the drag and heat flux but its location has to be determined carefully. Mehta et al. [6] have numerically studied the pressure oscillations around a spiked body. At the junction of shear boundary layer and the reattachment zone, a peak in the value of mean pressure is noticed. VirenMenzes et al. [7] have studied the efficiency of aerospike assemblies as drag reducers for blunt cones. The unsteady flow field around the spiked blunt cone and the increase in drag force on the body with a decrease in spike length are the observations made. Wood [8] has studied the factors which influence the shape as well as size of the steady separated flow region. This investigation has shown that the flow near the reattachment point mainly influences the size and shape of the region of separated flow. A numerical analysis to study the effects of adding a spike has been performed by Benjamin Meyer et al. [9], and it is concluded that wave and skin-friction drag, and heat transfer rates are observed to be characterized by the stand-off distance and also the ratio of diameter of blunt body to that of the spike. The drag reduction in spherical spike in front of blunt body has been investigated by Kamyar Mansour et al. [10]. Upon investigation, it is found that high stagnation pressure is the cause for drag exerted on the body. A transient study by Rajesh et al. [11] reveals that at Mach 6.1 and zero angle of attack, the frequency of the hypersonic buzz phenomenon reaches a peak value at critical *L/D* ratio. Transient investigation carried out by Rajesh et al. [12] at Mach 6.1 and zero angle of attack reveals that the frequency of the hypersonic buzz decreases proportional to corresponding rise in *L/D* ratio and at an *L/D* ratio of 2.5 K. Nagashetty et al. [13] have shed light on a method used in visualization of the shock, and the results observed that the visualized strong shock wave (detached) ahead of the body using this technique resembles the predicted shock wave with the same stand-off distance. Research has been done using the flow over spiked pointless bodies by Bogdonoff et al. [14] and observed that the order magnitude of pressure level and fraction of heat transfer rate calculated on spikeless hemisphere has been decreased by the use of a spike protruding from a hemispherical-nosed cylinder at $M = 14$. The alteration in time-averaged value of $C_d$ with surge in L/D ratios of spikes was analyzed by Rajesh et al. [15]. It is inferred that the foreshock–aftershock interaction point shifts away from the longitudinal axis with a rise in the L/D ratio, which decreases the $C_d$. An investigation about the flow over axisymmetric spiked bodies at a Mach number of 6.8 has been made by Maull et al. [16], and it shows that in some cases rapid fluctuation may occur due to the flow pattern. Consequences of numerical simulation of flow field over a spiked blunt nose are observed in this paper by Masafumi Yamauchi et al. [17]. The drag and separated region area is determined by the spike length. In the separated region, the drag reduction is caused by low-level pressure.

From the study of literature review, it can be observed that, the addition of a spike at the nose of a blunt body traveling at supersonic speeds reduces the drag force to a good extent. The drag reduction depends on the length, diameter and the shape of the spike and the angle of attack for a given Mach number. The common shapes are cut spike, flat aerodisk and hemispherical. On the contrary, addition of a spike also increases the static temperature of the spike. So, studying the characteristics related to spiked body at different Mach numbers is essential.

## 3 Geometry and Grid Generation

The geometry of the body that is investigated is given in Fig. 1a. A study by Menezes et al.[7] is also done on a similar body. Different spike shapes are shown in Fig. 1. These spike shapes have also been studied by Mehta et al. [9]. From Fig. 1a, *L* represents the total length of the spike and *D* represents the diameter of the body following the spike. The spike diameter *d* is taken as 0.02D, with *D* being the diameter of the cylindrical portion of the body. *L/D* ratio can vary up to 4, but here, an *L/D* of 1 is chosen as it offers better heat load, according to Barzegar et al. [1]. The fillet radius of the nose of body is taken as *D*/4, and the inclination of nose following the spike is at an angle of 30°, with respect to the vertical, i.e.,

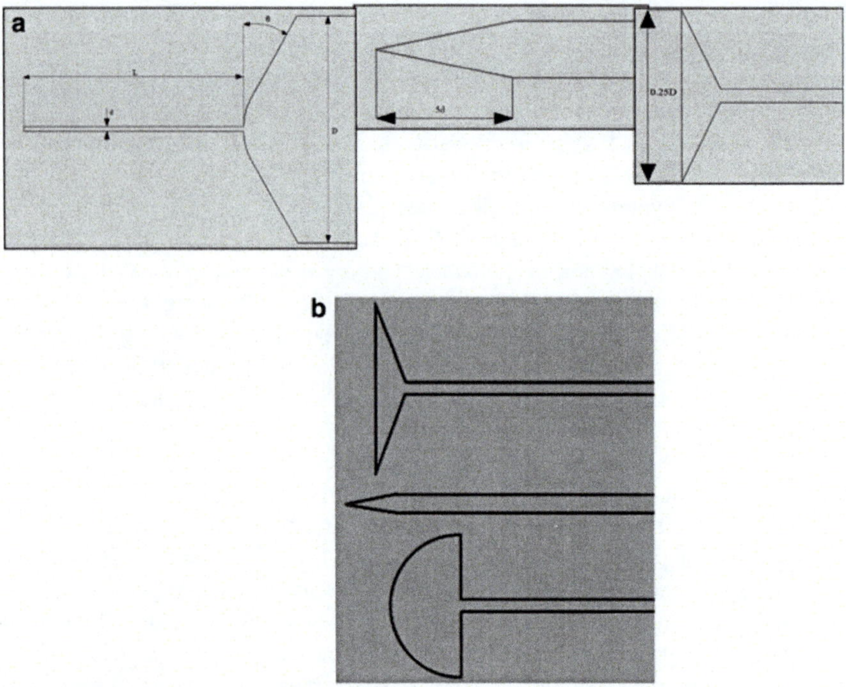

**Fig. 1** **a** Geometry of the body used for analysis and respective dimensions of sharp spike and aerodisk. **b** Different spike geometries used for analysis

perpendicular to the axis of the spike. The diameter of the spike for all the given configurations is constant, with the diameter of the aerodisk taken as $D/4$.

The mesh that has been used for investigation is structured mesh, with an element size of 0.5 mm. Quadrilateral elements are used as they offer better distribution thought the surface with lesser number of elements when compared to triangular elements. Moreover, the distribution of the elements can be achieved in such a way that the elements are parallel to the free flowing fluid, thereby increasing accuracy of results. Quadrilateral elements allow for increased stretching when compared to triangular elements. The grid is sized uniformly thought the surface. The quality and fineness of the grid play a major role in the solution accuracy. Hence, the generation of grid has been done with caution (Fig. 2).

## 4 Methodology

Since the whole problem is considered as an axissymmetric one with zero angle of attack, the whole domain is reduced to a half of the original domain for simpler calculations. The whole flow is considered as steady-state flow converging with

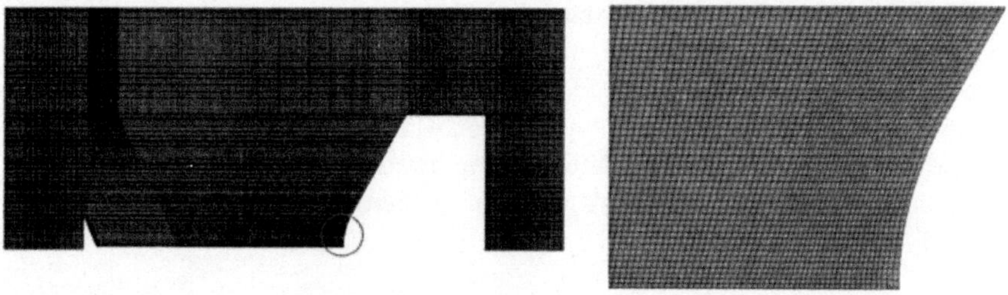

**Fig. 2** Grid generated for flat aerodisk and magnified at curved portion of body

absolute velocity formation and density based as the flow is compressible traveling at high Mach numbers. The viscosity is not established as a constant, but is varied with variation in temperature according to Sutherland law of viscosity.

A turbulence model of $K$-$\omega$ (SST) is preferred as it is a compromise between $K$-$\omega$ (standard), which yields solutions with good accuracy near the boundary, and $k$-$\varepsilon$, which yields solutions with good accuracy away from the boundary. Use of $K$-$\omega$ (SST) improves the predictive accuracy for flows with adverse pressure gradients. A finite volume discretization technique is employed to the flow field. Roe's approximate Riemann solver is used. Implicit formulation is used as it is more stable and accounts for parameters dependency also. Second-order flux limiter minimum model is used to increase spatial accuracy.

### *4.1 Boundary Conditions*

The domains of inlet, outlet and the far field are all assumed to be pressure far field, with the left most face being inlet and right most face being outlet. The wall and axis are assigned appropriately. The rest is assumed to be pressure far field. The nose temperature of the body is considered to be 300 K, with no-slip wall conditions. The far-field conditions are assumed as 425 Pa, and 140 K, with Mach number varying from 4.75 to 6.75 in steps of 0.5.

## 5 Results and Discussions

### *5.1 Validation*

Validation has been done for flat aerodisk at Mach 5.75 using ANSYS software. Perpetual ANSYS Academic Edition License is held by the affiliated university, and the results have been compared with that of Barzegar Gerdroobary et al. [1]. Similar Mach contours for the same Mach numbers have been obtained. Also, the density and velocity contours plotted by the author [1] show good matchings. Results of the first investigation show that further analysis can be continued and will yield acceptable and accurate results. Figure 3 shows the obtained Mach contour of a flat aerodisk at free flow Mach of 5.75. The recirculation zone above the spike can be observed in this contour. A streamlines contour (Fig. 4) has also been plotted in this case so as to verify if the recirculation region has been captured properly, because the change in drag mainly depends upon the properties of the fluid in this region.

## 5.2 Results

The simulations have been performed for Mach numbers from 4.75 to 6.75, varying in steps of 0.5. The relationship between the Mach numbers and drag coefficients ($C_d$) for various spike geometries has been analyzed. The results for various spike configurations have been compared, and the corresponding contours of Mach number and density have been shown.

Simulation shows the presence of a strong bow shock standing at a small distance in front of the spike. It can clearly be understood from the density contours as shown in Fig. 5a, b, the reason for causing an increased drag force on the spike tip. The point of reattachment of the main shock with the shoulder of the body can also be realized by plotting contours of respective configuration. Plotting a respective temperature contour, as in Fig. 5c, shows the region of reattachment of generated shock with the shoulder of body.

The flow induced by the aero spike separates due to pressure rise in front of the nose, exceeding the maximum allowable pressure for a boundary layer with attached flow. This causes a zone of recirculation, as can be seen in Fig. 6a. This area of recirculation depends upon the Reynolds number of the free flow, configuration of the spike, as well as the length of the spike.

It can also be seen in Fig. 6a that the streamline representing the shear layer along the surface does not touch the body, but changes direction and proceeds to the tip of the blunt nose.

**Fig. 3** Mach contour of flat aerodisk at Mach 5.75

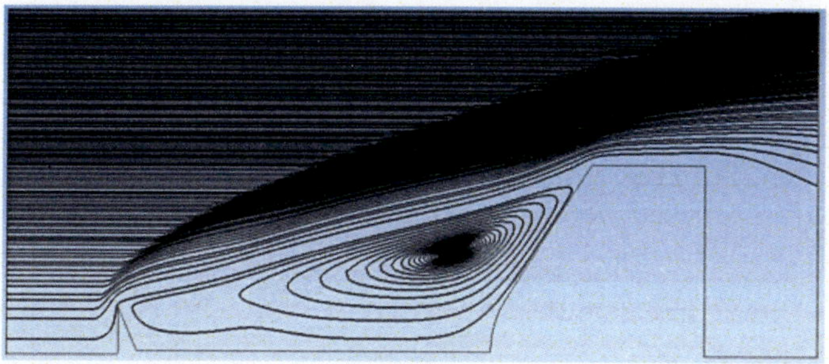

**Fig. 4** Streamlines of flat aerodisk at Mach 5.75

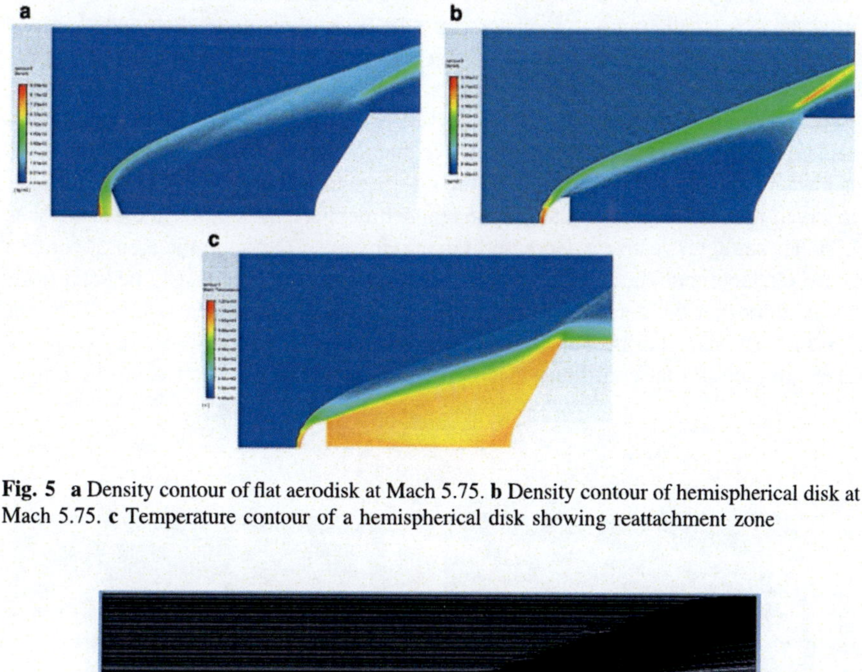

**Fig. 5** **a** Density contour of flat aerodisk at Mach 5.75. **b** Density contour of hemispherical disk at Mach 5.75. **c** Temperature contour of a hemispherical disk showing reattachment zone

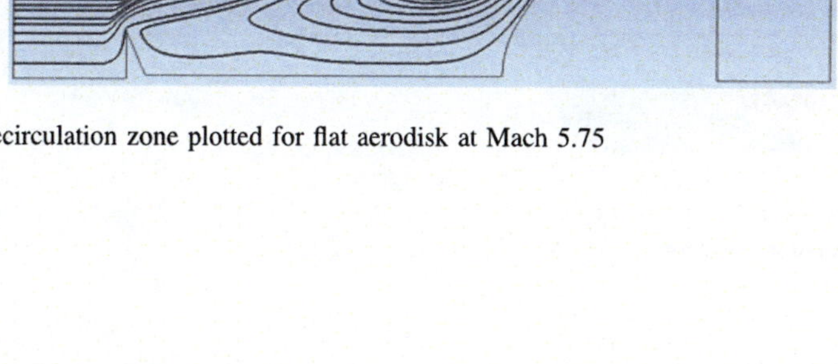

**Fig. 6** Recirculation zone plotted for flat aerodisk at Mach 5.75

This recirculation region also causes a deflection in the free flow, causing a secondary shock. The secondary shock generated can interfere with the primary shock and reattachment shock. There is also chance for another shock to take place after the recirculation zone, at the nose of the blunt body, yet again changing the drag force on the body. Contours for different spike configuration have been plotted, and relation between Mach number and $C_d$ has been found out in each of the cases.

## 5.3 *Analysis of Drag Coefficient for Cut Spike*

The cut spike is one of the most basic spike configurations. A bow shock can be observed at a small stand-off distance from the tip of the spike, when one observes the contours for this configuration. It can also be seen that another shock is generated at the point of reattachment at the shoulder of the body, from the contour. Also, the recirculation area can be observed from the contours for this configuration. Comparing Fig. 7a, b, we can observe that the density in the recirculation zone in front of the body is greater for Mach 4.75 than at Mach 6.75. It can also be observed that the area of contact of the reattachment shock is more in case of Mach 4.75. These factors cause a decrease in drag. However, although the area of contact of the reattachment shock is less in case of Mach 6.75, it can clearly be seen from this contour that the density is more in this region than that at Mach 4.75. This can cause an increase in drag. This is the reason why the decrease in $C_d$ is very small, as there are factors that contribute in drag reduction, as well as increasing the drag.

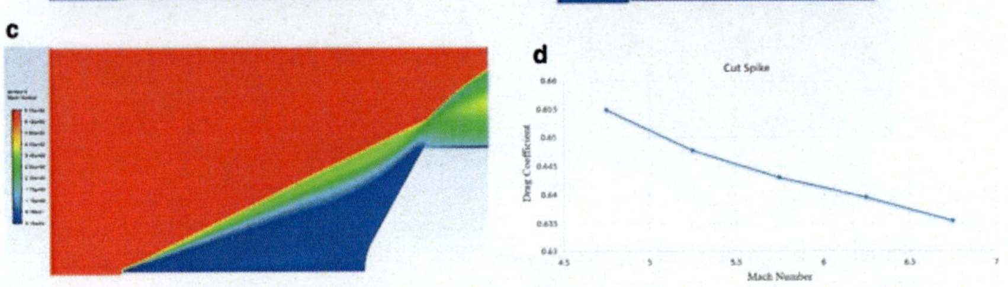


**Fig. 7** **a** Density contour of cut spike at Mach 4.75. **b** Density contour of cut spike at Mach 6.75. **c** Mach contour of cut spike at Mach 5.75. **d** Mach number versus $C_d$ graph for cut spike

**Table 1** Relationship between Mach number and $C_d$ for cut spike

| Mach number | Drag |
|---|---|
| 4.75 | 0.6549 |
| 5.25 | 0.6477 |
| 5.75 | 0.6430 |
| 6.25 | 0.6395 |
| 6.75 | 0.6354 |

Hence, it can be concluded that this configuration in not one of the best spike configurations to reduce drag. Mach contour of this configuration is plotted in Fig. 7c (Table 1).

Comparing $C_d$ values for this configuration from Mach 4.75 to 6.75, it can be observed that there is only a 2.97% decrease in drag coefficient, across the respective Mach numbers.

## 5.4 *Analysis of Drag Coefficient for Flat Aerodisk*

This spike configuration is a modification of the basic cut spike configuration. As can be seen from the figures pertaining to flat aerodisk, there is an increase in area at the spike tip, allowing an increase in the area for stagnation. Observing the contours for this configuration, we can see that flat aerodisk has a much better drag reduction property than a basic cut spike. This can be observed from the respective contours. Streamlines for this configuration are already shown in Fig. 6a, b. Separation zone and reattachment zone can clearly be observed in this geometry using respective contours. Reattachment point is more or less near the same region for flat aerodisk, when compared with cut spike at the respective Mach numbers. Comparing the flat aerodisk with cut spike, the area of reattachment is very less for this configuration. This reasons for a good drag reduction for this configuration. Comparing Fig. 8a, b, the density in front of the body is also lesser in case of Mach 6.75, when compared at Mach 4.75. Also, due to the increased area in front of the spike and the density being more at Mach 4.75 in front of the spike, an increase in drag can be observed at this Mach number. Hence, there is a significant decrease in drag coefficient, as one observes from Mach 4.75 to Mach 6.75. This can be seen from the contours plotted in Fig. 8a, b. A Mach contour has also been plotted in Fig. 8c to show the shock shape for this configuration (Table 2).

It can be observed that there is 13.26% decrease in drag coefficient as one moves from Mach 4.75 to Mach 6.75.

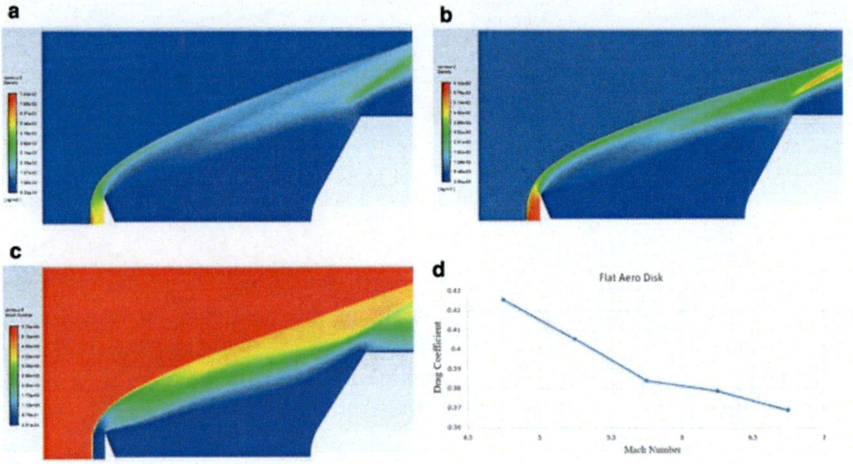


**Fig. 8** **a** Density contour of flat aerodisk at Mach 4.75. **b** Density contour of flat aerodisk at Mach 6.75. **c** Mach contour of flat aerodisk at Mach 5.75. **d** Mach number versus $C_d$ graph for flat aerodisk

**Table 2** Relationship between Mach number and $C_d$ for flat aerodisk

| Mach number | Drag coefficient |
|---|---|
| 4.75 | 0.4253 |
| 5.25 | 0.4051 |
| 5.75 | 0.3873 |
| 6.25 | 0.3786 |
| 6.75 | 0.3689 |

## *5.5 Analysis of Drag Coefficient for Hemispherical Aerodisk*

A modification of the flat aerodisk configuration gives hemispherical disk configuration. This configuration also shows good drag reduction properties when compared to cut spike configuration, but not when compared to flat aerodisk configuration. This can be observed by comparing both the configurations at the respective Mach numbers.

Density contour for this configuration has been plotted in Fig. 9a, b. At Mach 6.75 for this configuration, it can be seen that at the secondary shock generated at the shoulder of the body, there is an increased density, causing an increase in drag. This is not so during Mach 4.75. But we can also see that the density in front of the nose is significantly less in case of Mach 6.75. This results in a significant decrease in drag, when compared to that at Mach 4.75. Also, the area in front of the spike is more, when compared to cut spike, with the density distribution being more or less same in this region at both the Mach numbers. Hence, it can be observed that an important factor affecting the drag force for this configuration with an increase in

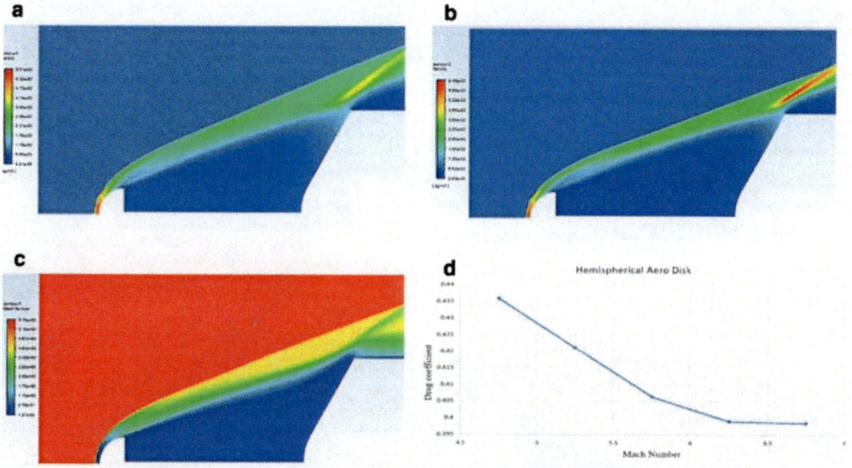


**Fig. 9** **a** Density contour for hemispherical disk at Mach 4.75. **b** Density contour for hemispherical disk at Mach 6.75. **c** Mach contour for hemispherical disk at Mach 5.75. **d** Mach number versus $C_d$ graph for hemispherical aerodisk

**Table 3** Relationship between Mach number and $C_d$ for hemispherical aerodisk

| Mach number | Drag coefficient |
|---|---|
| 4.75 | 0.4358 |
| 5.25 | 0.4211 |
| 5.75 | 0.4063 |
| 6.25 | 0.3995 |
| 6.75 | 0.3984 |

Mach number is the density difference in front of the nose, that is in the recirculation area. This decrease in drag outweighs the increase in drag at the shoulder of the body (Table 3).

As one moves from Mach 4.75 to Mach 6.75, nearly 8.5% reduction in the $C_d$ value can be seen.

### 5.6 *Analysis of Drag Coefficient for Sharp Spike*

When compared to the previous configurations, sharp spike has a high $C_d$ value for the same Mach numbers. As we can observe from Fig. 10a, b, the density distribution in front of the nose of the body is more or less similar. But in case of reattachment at the shoulder, significantly more area is in contact in case of Mach 4.75, as can be seen from the contour. It can be observed that the density distribution at the shoulder of the body is more at Mach 6.75, causing an increase in

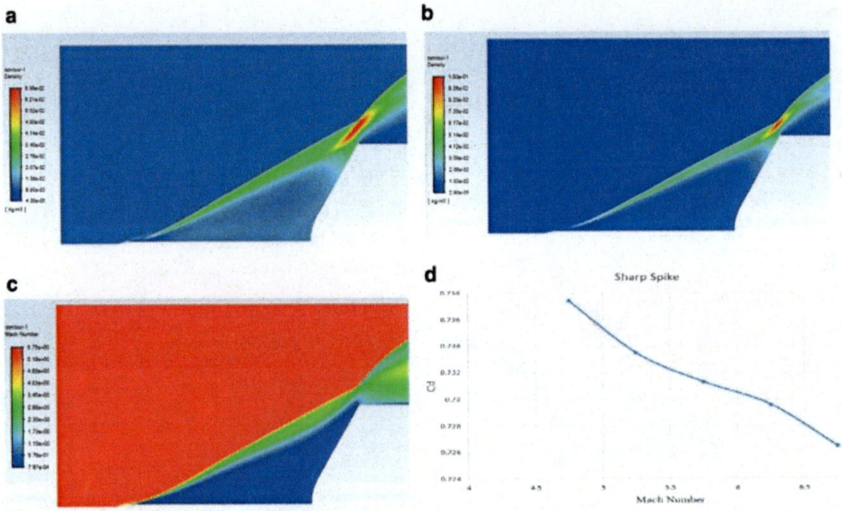


**Fig. 10** **a** Density contour for sharp spike at Mach 4.75. **b** Density contour for sharp spike at Mach 6.75. **c** Mach contour for sharp spike at Mach 5.75. **d** Mach number versus $C_d$ graph for sharp spike.

**Table 4** Relationship between Mach number and $C_d$ for sharp spike

| Mach number | Drag coefficient |
|---|---|
| 4.75 | 0.7375 |
| 5.25 | 0.7334 |
| 5.75 | 0.7313 |
| 6.25 | 0.7299 |
| 6.75 | 0.7265 |

drag, the area in contact of the reattachment decreases, decreasing the drag in this case. Both the factors combined cause a slight decrease in the drag for this configuration, as one observes from Mach 4.75 to Mach 6.75. Also, when compared to other configurations, huge amount of heat is generated on the body in case of sharp spike. The Mach contour for this configuration is shown in Fig. 10c (Table 4).

It is observed that $C_d$ value decreases by about 1.49% with an increase in flow velocity from Mach 4.75 to Mach 6.75.

### 5.7 *Comparison Across Different Spike Configurations at Different Mach Numbers*

Also, comparisons have been made between different configurations for the same Mach number of 4.75, 5.75, and 6.75. When comparing sharp spike and cut spike at

**Table 5** Relationship between various spike geometries and $C_d$ at Mach 4.75

| Mach number | Cut spike | Flat aerodisk | Hemispherical aerodisk | Sharp spike |
|---|---|---|---|---|
| Drag coefficient | 0.6549 | 0.4253 | 0.4358 | 0.7375 |

**Table 6** Relationship between various spike geometries and $C_d$ at Mach 5.75

| Mach number | Cut spike | Flat aerodisk | Hemispherical aerodisk | Sharp spike |
|---|---|---|---|---|
| Drag coefficient | 0.6430 | 0.3873 | 0.4063 | 0.7313 |

**Table 7** Relationship between various spike geometries and $C_d$ at Mach 6.75

| Mach number | Cut spike | Flat aerodisk | Hemispherical aerodisk | Sharp spike |
|---|---|---|---|---|
| Drag coefficient | 0.6354 | 0.3689 | 0.3984 | 0.7265 |

Mach 4.75, from the density contours, we can observe the generation of a strong secondary shock at the shoulder of the body, rather than in front of the spike tip, as can be seen in Figs. 7a and 10a, increasing drag, showing that they are not efficient physical drag reducing configurations. At the same Mach number, simulations yield better results for flat aerodisk and hemispherical disk configurations. The drag coefficients for various spike configurations at Mach numbers 4.75 and 6.75 are shown in Tables 5, 6, and 7, respectively.

From the results in Tables 5, 6, and 7, it can be concluded that as there is an increase in Mach number, a decrease in drag coefficient occurs. About 42.33%, 49.55%, 49.23% decrease in drag coefficients can be observed at the same Mach numbers of 4.75, 5.75, and 6.75, respectively, comparing flat aerodisk and sharp spike. Also, from Fig. 11, we can see that the drag reduction of both the aerodisks is similar, as they have more contact area in front of the spike, and less reattachment area near the shoulder of the body, as not in the case of the other configurations.

Maximum drag can be observed at Mach number 1 and gradually decreases with increase in Mach number. This is because, at Mach 1, there are still regions in the body where the flow returns to subsonic flow, after passing through the shock. An abrupt reduction in speed and a corresponding increase in pressure will take place, causing an increase in net drag.

As the Mach number gradually increases, the flow slowly starts to become supersonic everywhere (except for small regions where stagnation occurs). The bow shock generated does not come in contact with the boundary layer, and drag as compared to lesser Mach number is not observed.

A graph plotted between Mach number (along horizontal axis) and drag coefficient (along vertical axis) is shown in Fig. 11. It is evident from this graph that spikes with aerodisk (flat and hemispherical) geometries reduce drag more than cut and sharp spike. A linear decrease in the respective drag coefficients can be observed with increase in Mach number. Also, from this graph, one can observe

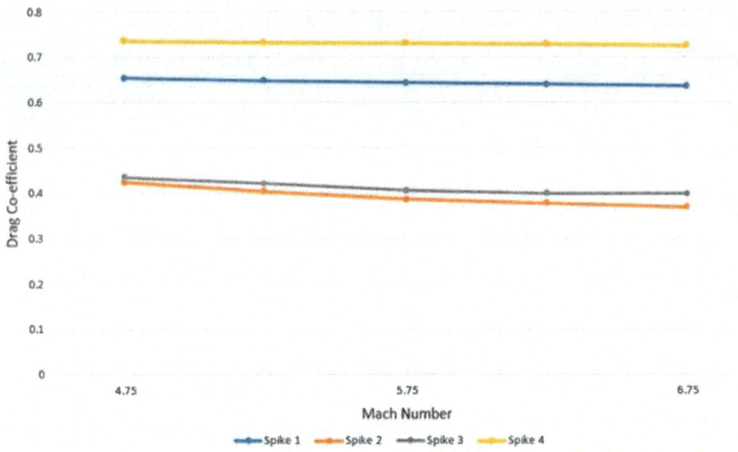


**Fig. 11** $C_d$ for various spike geometries at various Mach numbers

that aerodisk configurations tend to decrease drag more with increase in Mach number. Hence, use of geometries like flat aerodisk and hemispherical aerodisk is more preferred over the other configurations discussed.

In the respective graph,
Spike 1: Cut spike configuration.
Spike 2: Flat aerodisk configuration.
Spike 3: Hemispherical aerodisk.
Spike 4: Sharp spike configuration.

# 6 Conclusion

- The numerical investigation carried out on the four spike geometries in 2D steady state yielded that a blunt spike with $L/D$ ratio = 1 induces a conical detached bow shock in front of nose of the body.
- The size of recirculation zone in front of the nose mainly depends upon the free stream Reynolds number, length of the spike, and geometry of the spike tip, i.e., the configuration of the spike.
- A decrease in drag coefficient can be observed as Mach number of free stream increases, with the maximum observed drag coefficient occurring at Mach 4.75, and gradually decreasing with increase in Mach number.
- This is observed because an abrupt decrease in velocity causes a corresponding increase in pressure, which in turn causes an abrupt increase in the resultant drag.

- It has been concluded that aerodisk configurations causes lesser drag, when studied at the same Mach number.
- Point of reattachment and density change in the recirculation zone play a major role in influencing drag of the induced shock with the main body.

## References


1. BarzegarGerdroodbary M, Hosseinalipour SM (2010) Numerical simulation of hypersonic flow over highly blunted cones with spike. Acta Astronaut 67(1–2):180–193
2. Miliçev SS, Pavloviç MD (2012) Influence of spike shape at supersonic flow past blunt-nosed bodies: experimental study. AIAA J 40(5):1018–1020 (technical notes)
3. Mehta RC (2010) Numerical simulation of the flow field over conical, disc and flat spiked body at Mach 6. Aeronaut J 114(1154):225–236
4. Gopalan J, Menezes V, Reddy KPJ, Hashimoto T, Sun M, Saito T, Takayama K (2005) Flow fields of a large-angle, spiked blunt cone at hypersonic Mach numbers. Trans Japan Soc Aeronaut Space Sci 48(160):110–116
5. Sreekanth N, Akhil J, Nagaraja SR (2016) Design and analysis of secondary spike on blunt head. Indian J Sci Technol 9(43). https://doi.org/10.17485/ijst/2016/v9i43/104649
6. Mehta RC (2000) Numerical heat transfer study over spiked blunt bodies at Mach 6.8. J Spacecraft Rockets 37(5):700–703
7. Menezes V, Saravanan S, Jagadeesh G, Reddy KPJ (2003) Experimental investigations of hypersonic flow over highly blunted cones with aero-spikes. AIAA J 41(10):1955–1966
8. Wood CJ (1962) Hypersonic flow over spiked cones. J Fluid Mech 12(Pt. 4):614–627
9. Meyer B, Nelson HF, Riggins DW (2001) Hypersonic drag and heat transfer reduction using a forward-facing jet. J Aircr 38(4):680–686
10. Mansour K, Khorsandi M (2014) The drag reduction in spherical spiked blunt body. Acta Astronaut 99:92–98
11. Rajesh R, Rakesh SG (2018) Effect of dimensions of various spiked cylinders on the buzz phenomenon subjected to hypersonic flows. Int J Fluid Mech Res 45(5):377–388
12. Ramachandran R, Ganesh RS (2017) Effect of dimensions of sharp spiked cylinder on the buzz phenomenon subjected to hypersonic flows. Int J Fluid Mech Res 44(6):469–485
13. Nagashetty K, Syed Saifuddin K, Saravanan S, Gurumurthy KS, Jagadeesh G, Reddy KPJ (2000) Visualization of shock shapes around blunt bodies at hypersonic Mach number in a shock tunnel using electrical discharge technique. Curr Sci 79(8):1086–1089
14. Bogdonoff SM, Vas IE (1959) Preliminary investigations of spiked bodies at hypersonic speeds. J Aero/Space Sci 26(2):65–74
15. Rajesh R, Rakesh SG (2020) Effect on the drag coefficient of various spiked cylinders during buzz phenomenon subjected to hypersonic flows. J Brazilian Soc Mech Sci Eng 42(6), article number 288. https://doi.org/10.1007/s40430-020-02384-5
16. Maull DJ (1960) Hypersonic flow over axially symmetric spiked bodies. J Fluid Mech 8:584–592
17. Fujita M, Kubota H (1992) Numerical simulation of flow field over a spiked blunt nose. Comput Fluid Dyn J 1(2):187–195

# Experimental Investigations of Nano-$BaTiO_3$-Filled HTV SiR Insulating Specimen Under Rainy Conditions

**Vinayak V. Rao, K. Ramakrishna Murthy, G. M. Mamatha, R. Hari Krishna, and Pradipkumar Dixit**

**Abstract** Outdoor insulator forms an integral part of the power system. Its failure at normal working voltage due to flashover affects the reliability of the electrical network. Recent literature has shown that the addition of nanosized barium titanate ($BaTiO_3$) fillers increases the electrical insulating properties of the energy storage devices. This factor is extremely useful for the outdoor insulation, as it helps in withstanding higher electrical stress improving the flashover voltage under wet conditions. In the present work, 1, 3, and 5 wt% nanosized barium titanate-filled silicone rubber specimens were prepared. The barium titanate was synthesized in the laboratory using co-precipitation method. X-ray diffraction test was done to ascertain the particle size of the filler is found to be 68.5 nm which is the nanorange. Contact angle measurements revealed that 5 wt% nanofilled SiR has 97° which indicates better hydrophobicity. Flashover voltages measured under dry and wet conditions show that the specimens with 5 wt% filler concentration have a value of 31.11 kV and 21.98 kV, respectively, which is better as compared to the unfilled, 1 wt% and 3 wt% filled nanocomposites. Fourier transform infrared spectroscopy results showed that 5 wt% specimens have higher absorbance values at 1150 wavenumber depicting that it has strong Si–O–Si bond. The higher value of absorption at 1250 wavenumber shows increased content of methyl groups in Si–$CH_3$ bond indicating better hydrophobicity. Strong Si–O–Si bond, superior hydrophobicity, and positive temperature coefficient of resistance of nanofillers are the prime factors responsible for the improvement in flashover voltage of barium titanate-filled nanocomposites. This makes the use of these nanofilled composites ideal for outdoor insulation under rainy conditions.

**Keywords** Barium titanate · Hydrophobicity · Flashover voltage · Nanocomposites · Polymeric insulator

V. V. Rao (✉) · K. Ramakrishna Murthy · G. M. Mamatha · P. Dixit
Department of Electrical and Electronics Engineering, M.S. Ramaiah Institute of Technology, Bangalore 560054, India

R. Hari Krishna
Department of Chemistry, M.S. Ramaiah Institute of Technology, Bangalore 560054, India



R. Kumar et al. (eds.), *Recent Trends in Thermal Engineering*, Lecture Notes in Mechanical Engineering, https://doi.org/10.1007/978-981-16-3132-0_16

# 1 Introduction

Outdoor insulators are integral part of the power system. Its performance is a key factor which effects the reliability of the electrical networks. Traditional ceramic or glass insulators are being replaced by the polymeric insulators due to numerous advantages. In the long term, these insulators undergoes surface degradation, reducing the hydrophobicity when subjected to environmental electrical and mechanical stresses. The environmental stresses are UV, humidity, rain, biological growth, electrical stresses are leakage current, dry band arcing, corona and mechanical stress are cyclic loadings. Among these various factors, rainy condition is an important factor as these insulators are exposed to it during monsoon months in major parts of India. Under these conditions, with decrease in surface resistance of the insulator, leakage current starts to flow, forming the dry bands. This further results in flashover of the insulators, ultimately failure under normal operating voltages. The power outages arising due to these situations will result in a huge economic losses [1].

The polymeric insulators performance under rainy conditions can be improved by incorporation of inorganic nanofillers into the polymer base. This will improve its electrical, mechanical, hydrophobicity, and thermal properties. The fillers size, shape, degree of adhesion, and orientation will have an effect on the polymer chain. These fillers can be added at micro/nanolevel with specific weight percentage in order to improve the properties of the insulator. But care should be taken to see that these fillers should be of optimum level as they tend to increase electrical property but reduce mechanical or hydrophobic property [2].

Barium titanate is one of the possible nanofillers into the polymer base as it possesses good electrical insulating properties. This material is being extremely used in electroceramic, electro-optical, and electromechanical applications. The reason for their usage is high dielectric strength, lead-free content which makes it environmental friendly, and positive temperature coefficient of resistance which makes it extremely useful in the field of capacitors. The $BaTiO_3$ can be used as semiconductor, ferromagnetic, piezoelectric, and pyroelectric materials. These are possible as it possesses large polarization, permittivity, and greater induced strain [3].

In the case of outdoor high voltage insulation system, reducing the electric stress becomes important factor as it impedes the dry band arcing and partial discharges. This can be achieved by using the high permittivity nanofillers like barium titanate which will improve the insulation of the system. Silicone rubber has been filled with nanobarium titanate with specific, shape, size, concentrations to minimize the material erosion which lowers the temperature at dry band sites. Results indicate that the electric field stress has reduced due increase in relative permittivity. This has caused due to the oxidative layer on metal particle filler and its dipole movements orientation which in turn improves the performance of the outdoor insulators [4].

Vulcanized natural rubber with ferroelectric composites is an alternative option to use in outdoor insulation. This material combination possesses high dielectric strength, better permittivity, and enhanced tensile strength. Among the ferroelectric ceramic family, $BaTiO_3$ filler is better as they possess most of the above-mentioned properties and they are environmental-friendly. In this regard, work was carried out by incorporating $BaTiO_3$ into the natural rubber. The mechanical, morphological, and electrical properties of the composites were explored by the author. Results show that the dielectric constant to increase with reduction in tensile strength. This factor was attributed to a suitable conduction mechanism in the composites at room temperature, which was observed due to quantum mechanical tunneling among the SMBT particles [5].

In outdoor insulation, silicone rubber filled with high dielectric constant materials is preferred, as it possesses low cost, lightweight, large strains, and high energy densities. In this category, $BaTiO_3$ is considered to be a best alternative as it reduces the electric field distribution due to its high permittivity. In order to achieve this objective, work was carried out by incorporation of micro-$BaTiO_3$ into the silicone rubber with weight percentage ranging from 10 to 50%. SEM results showed that there is uniform distribution of $BaTiO_3$ particles into the polymer base. The $BaTiO_3$ composites exhibited enhanced strain under applied electric field. Higher dielectric strength was observed with higher filler concentrations of $BaTiO_3$ [6].

The common phenomena occurring in the silicone rubber insulator used under contamination conditions are tracking and erosion. This results in a conductive path leading to flashover and failure of the insulator. In this regard, work has been carried out to enhance the tracking and erosion resistance of polymeric specimen by incorporation of nano-$BaTiO_3$ with a fraction ranging from 0.5, 1, and 2 wt%. The surface degradation study was carried out with inclined plane test setup. Weight loss and surface roughness values were found to be low in nanofilled specimens. SEM images reveal that there is a presence of dents in the unfilled specimen due to the leakage currents [7].

From the literature survey, it is evident that most of the researcher have carried out synthesis of $BaTiO_3$ and studied its electrical, mechanical, and structural properties. Only few authors have incorporated $BaTiO_3$ at micro/nanolevel into the silicone rubber specimen. Further, they have done investigations on the performance of these polymer composites. So none of the authors have attempted to synthesize and study the performance of $BaTiO_3$-filled polymers under rainy conditions. Since most of the literature indicates $BaTiO_3$ has excellent insulating property and environment-friendly, it is ideal filler for the outdoor insulation used in high voltage system. In this regard, the present study focuses on synthesis of nano-$BaTiO_3$ using co-precipitation method. The preparation of silicone rubber nanocomposites with 1, 3, and 5 wt% $BaTiO_3$ fillers. The characterization tests such as FTIR and contact angle were carried out on these specimens. Further experiments are carried out on these unfilled, nanofilled specimens under dry and wet conditions with test arrangement done in the laboratory.

## 2 Materials and Methods

In the present study, $BaTiO_3$ has been considered as it has high dielectric strength, good permittivity, and positive temperature coefficient of resistance and is lead free. Co-precipitation method is adopted for the preparation of nano-$BaTiO_3$ fillers.

Synthesis of the filler is as follows:

Nitric acid of 20 ml is taken in a Petri dish, is gradually added to 20 ml of distilled water, and this solution is stirred for 10 min. In a separate beaker, 10 ml of ethanol and 5 g of barium acetate is mixed with 6 ml of titanium tetraisopropoxide. Further, both the solutions are mixed together in a 50 ml distilled water. This obtained solution is stirred using magnetic stirrer for 3 h and centrifuged for 5 min. This mixture is calcinated at 600 °C for 3 h. The resultant powder is collected and stored in an airtight container.

In order to analyze the phase purity and the crystalline nature of the prepared sample, PXRD analysis was carried out and is presented in Fig. 1. From the diffraction pattern it was inferred that all the peaks readily match with the standard JCPDS card number 31–0174. This confirms the formation of $BaTiO_3$, and the high intense diffraction peaks suggests the crystalline nature of the product. In order to calculate its crystalline size, Scherrer's equation was adopted. Using Scherrer's equation, the crystallite size of $BaTiO_3$ is found to be $\sim$68 nm.

The nanocomposite polymeric specimen is prepared by mixing fumed silica, alumina trihydrate, and $BaTiO_3$ in 1, 3, and 5 wt% thoroughly using sigma kneader equipment. This equipment uses 2HP induction motor which runs at 600 rpm and has a mixing capacity of 1 kg. Specific quantity of the mixture from sigma kneader is loaded into rubber injection molding machine. The mixture is hot pressed at 5000 C for 20 min. The resultant specimen is then cooled until they become hard in nature.

In order to study its chemical bonding, surface hydrophobicity, flashover phenomena under dry and wet conditions, FTIR, contact angle, and flashover studies were performed. In the present work, FTIR was performed using Bruker ALPHA-II

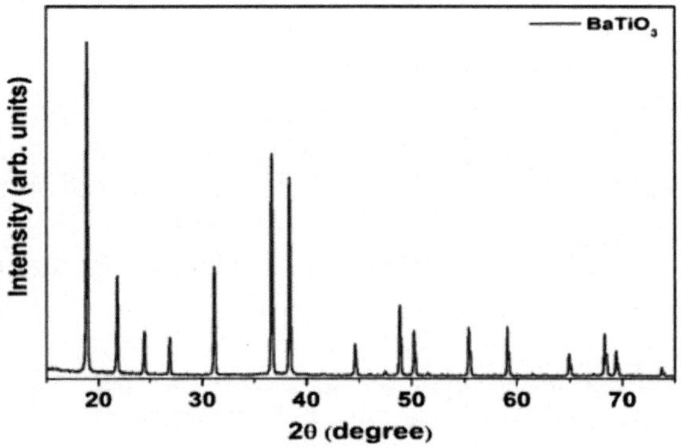


**Fig. 1** XRD plot for synthesized $BaTiO_3$

ATIR FTIR instrument is used and static hydrophobicity was measured using Image-J software and flashover studies under dry and wet condition on the polymeric insulating specimens when subjected to HVAC stress carried out according to IEC 383 [8].

### 2.1 *Experimental Setup for Conducting Dry and Wet Flashover Tests*

The composite specimen has been energized with 5 kVA, 230/50 kV, 4.5% impedance HV transformer. Its arrangement is shown in Fig. 2. The temperature and pressure of the laboratory during the experimentation have been recorded as 25–33 °C. The pressure is 981 mbar.

The flashover tests have been carried out in the fog chamber made up of acrylate (30 × 30 × 30 cm). The copper electrodes arrangement is provided inside the chamber with one end energized and the other is grounded. The experiments have been conducted as per IEC 60,583 [9].

## 3 Results and Discussions

### 3.1 *Dry Flashover Test*

The flashover trials were carried out on the composite molds with the facility described in the Sect. 2.1. Figure 3 shows average of five trials conducted on unfilled and nanofilled specimens.

From Fig. 3, it indicates under dry conditions, unfilled specimen has flashover voltage of 28.8 kV, 1% $BaTiO_3$ (29.14 kV), 3% $BaTiO_3$ (30.27 kV), and 5%

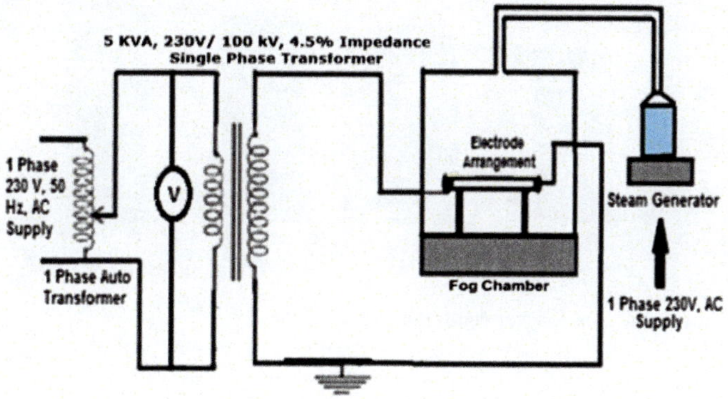


Fig. 2 Schematic diagram for flashover test setup

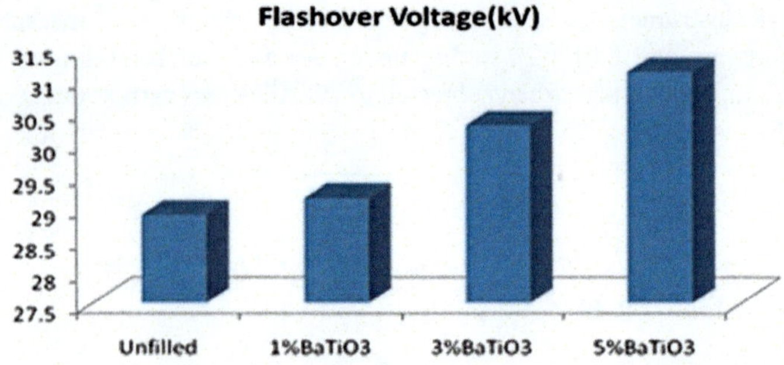


**Fig. 3** Flashover voltages for dry condition in kV

$BaTiO_3$ (31.11 kV). This means, 5% $BaTiO_3$ nanocomposite's performance is marginally higher than its counterparts.

## 3.2 *Wet Flashover Test*

The flashover tests were carried out on the specimen with relative humidity of 90–95% with the generation of steam inside the chamber. Figure 4 shows the average wet flashover voltage of five trials conducted on unfilled and nanofilled specimens.

From Fig. 4, it indicates under wet conditions, unfilled specimen has flashover voltage of 17.09 kV, 1% $BaTiO_3$ (19.52 kV), 3% $BaTiO_3$ (20.56 kV), and 5% $BaTiO_3$ (21.98 kV). This means, 5% $BaTiO_3$ performance is better than other specimens and is well suited for rainy conditions.

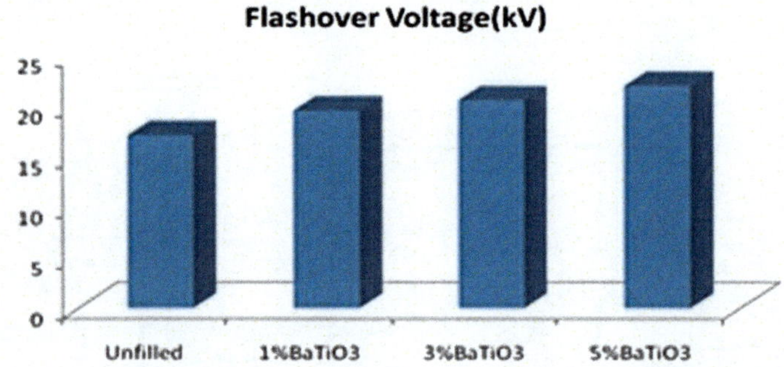


**Fig. 4** Flashover voltages for wet condition in kV

## 3.3 Contact Angle

The image J software has been used to measure the contact angle of the polymeric specimen. The obtained values are 92° in the case of unfilled specimens and in the case of 1 wt% $BaTiO_3$ nanocomposite 94°, 3 wt% $BaTiO_3$ is 95° and 5 wt% $BatiO_3$ it is 97° indicating nanofilled specimens have considerably better hydrophobicity.

## 3.4 Fourier Transform Infrared Spectroscopy (FTIR)

The severity in material degradation can be accessed by difference in infrared absorbance spectra. Presence of Si–O–Si bonds suggests the stability of the specimen. In the present case, absorption bands in the range of 1000–1100 $cm^{-1}$ can be attributed to the Si–O–Si bond in the silicone rubber. Appearance of these bands corresponding to Si–O–Si suggests that specimens are resistance to degradation. Absorption bands in the wavenumber range 1255–1270 are due to Si–$CH_3$ bond and it depicts hydrophobicity property of the material. Figure 5 shows FTIR graph for the all the specimen.

In order to analyze its resistance to bond breaking or scission, the absorbance values are recorded at the wavenumber 1100 $cm^{-1}$, and it is found that unfilled

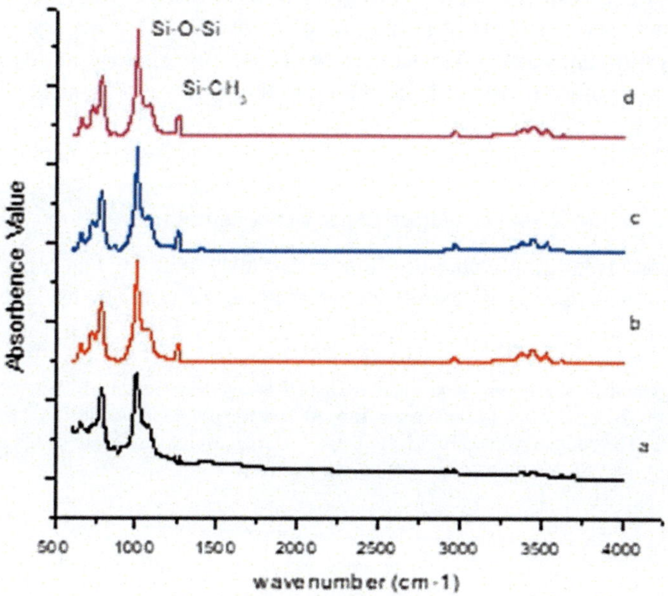


**Fig. 5** FTIR graph (a) unfilled specimen, (b) 1% $BFaTiO_3$, (c) 3% $BaTiO_3$, (d) 5% $BaTiO_3$

specimen has least value of 1.22, 1% $BaTiO_3$ (1.28), 3% $BaTiO_3$ (1.32), and 5% $BaTiO_3$ (1.34). This indicates 5% $BaTiO_3$ has better resistance to degradation due to strong Si–O–Si bond.

The percentage variation in absorbance values are noted for the wavenumber 1250 Si–$CH_3$, and it is found that unfilled specimen has least value of 0.20, 1% $BaTiO_3$ (0.21), 3% $BaTiO_3$ (0.22) and 5% $BaTiO_3$ (0.24). In 5 wt%, 5% $BaTiO_3$ value is marginally higher due to increase in $CH_3$ functional groups in silicone backbone. This has improved the surface hydrophobicity, due to the migration of low molecular weight from the polymer base to the surface.

## 4 Conclusions

The present work discusses synthesis of nano-$BaTiO_3$ and polymer nanocomposites specimens incorporating this fillers in 1, 3, and 5 wt%. XRD test reveals the synthesized $BaTiO_3$ crystal size is 69.5 nm which is in the supernanorange. The following inference was drown from the experimental work.

- The flashover voltages of unfilled and nanofilled specimens under dry and wet conditions show that 5 wt% nanocomposite values are marginally higher than that of other specimens.
- Contact angle results show better hydrophobicity was found 5 wt% $BaTiO_3$ specimens.
- The absorption values from the FTIR graph shows resistance to degradation, and hydrophobicity was superior in the case of 5 wt% $BaTiO_3$ category.
- $BaTiO_3$ filler has positive temperature coefficient of resistance means its resistance increases with rise in temperature as this case prevails under flashover conditions.
- Strong Si–O–Si bond, better hydrophobicity, and positive temperature coefficient resistance property of $BaTiO_3$ are the prime factors attributes to the improvement of flashover voltage under wet conditions.

One more advantage of using this filler is that it is environmental-friendly. This makes nanofilled $BaTiO_3$ as one of the alternatives for outdoor insulation under rainy conditions.

**Acknowledgements** The authors would like to thank Management, Principal, Ramaiah Institute of Technology, Bangalore for their constant support and for providing facility to carry out the research work. The authors would also like to thank Mr. Raghavendra Shanbag of Ms Innovative Silicone, Bangalore for his support in carrying out the work.

## References

1. Amin M, Salman M (2006) Aging of polymeric insulators (an overview). Rev Adv Mater Sci 92–136
2. Momen G, Farzaneh M (2011) Survey of micro or nano filler use to improve silicone rubber for outdoor insulators. Rev Adv Mater Sci 1–13
3. Ertug B (2013) The overview of the electrical properties of barium titanate. Am J Eng Res 2(8): 1–7
4. Cherney EA (2005) Silicone rubber dielectrics modified by inorganic fillers for outdoor high voltage insulation applications. IEEE Trans. Dielectr Electr Insul 12(6)
5. El-Gamal (2019) Electrical and mechanical characterization of vulcanized natural rubber filled with $BaTiO_3$ ceramic filler. J Thermoplast Compos Mater 1–16
6. Jiang L, Betts A, Kennedy D, Jerrams S (2015) The fabrication of dielectric elastomers from silicone rubber and barium titanate: employing equi-biaxial pre-stretch to achieve large deformations. Technological University Dublin, Technical Article
7. Kumar R, Gupta N (2015) Tracking and surface degradation of barium titanate filled silicone rubber nanocomposites. In: IEEE conference on electrical insulation and dielectric phenomena (CEIDP), Ann Arbor, MI, USA
8. Insulator string and insulator set for ac system—definition, test methods and acceptance criteria, IEC 383:1993 (1993)
9. Electrical strength of insulating materials—test methods-Part1 test at power frequency—IEC 60243 (2013)

# Exergy Analysis of a 660 MW Thermal Power Plant

**Keval Nikam, Ravinder Kumar, and Ravindra Jilte**

**Abstract** The energy demand is increasing exponentially day by day. It has become compulsory to upgrade the thermal equipment to reduce energy losses in quality terms. Most of the existing power plant has been designed according to energy analysis, where quality of energy was neglected. Second law thermodynamic focused on quality of energy which leads to exergy analysis. The present paper deals with exergy analysis of 660 MW coal-fired power plant using excel spreadsheet approach. The Xsteam solver pulgin is utilized in excel to evaluate the specific enthalpy, specific entropy at various location. The articles also provide the formulation to find the exergetic efficiency of components involved in plant. Overall, second law efficiency of the plant has been evaluated as 35.28%. Condenser contributes to low exergetic efficiency of 34.91%. The validation of the excel sheet model approach is done with the available literature. The present paper reveals that the excel sheet approach is highly accurate and less time-consuming method to evaluate exergy at different points in the system. This study will also help the researcher to develop such a simplified strategy in other thermal sectors.

**Keywords** Exergy · Spreadsheet · Exegetic efficiency · Condenser

## 1 Introduction

The energy consumption per person drastically has increased in twenty-first century. Because of this, it is necessary to develop thermodynamic model of shortly commence thermal power plant. The country has adapted strategy to build and convert existing power plant into supercritical power plant of 660 MW capacity [1]. The thermodynamic analysis is performed using thermodynamic first law where the energy and mass balance formulations are used to tabulate the losses. The limitation

K. Nikam · R. Kumar (✉) · R. Jilte
School of Mechanical Engineering, Lovely Professional University, Phagwara, Punjab 144411, India
e-mail: ravinder.22218@lpu.co.in

 
R. Kumar et al. (eds.), *Recent Trends in Thermal Engineering*, Lecture Notes in Mechanical Engineering, https://doi.org/10.1007/978-981-16-3132-0_17

of energy analysis could not identify the energy degradation in a thermodynamic process and the factor accountable for low plant efficiency. So, exergy analysis overcomes the limitation of energy analysis. This method helps in designing equipment utilization and strategies for using energy to maximum extent in newly installed plants. Hence, it is a powerful tool for measurement of energy quality, thereby helps to make the complex thermodynamic system more efficient. Some previous attempts were made to predict thermodynamic performance in terms of energy and exergy related to subcritical power plant in India [2–4]. Some researchers studied uncertainties in constant entropy process and exergetic efficiency in high-pressure turbine power plant resulted that dependency of uncertainty parameters are more rely on pressure as compared to temperature [5]. Early studies show some exergy analysis on subcritical thermal power plant assisted with solar feedwater heating unit found to be a better option than installed plant [6]. Exergy analysis was performed to study the effect on the efficiency due to steam exhaust pressure, temperature and pressure of incoming water, variable load in supercritical plant [7]. Some engineers also identified exergy destruction components, namely water wall, screen heater, primary heater which are different components than conventional power plant [8]. Many researchers have used the excel sheet approach as a tool for exergy analysis of various components present in the steam power plant [9, 10]. Indirectly exergy analysis implies how nearly the systems operation approaches the ideal conditions. The solver of XSteam was used to evaluate the properties of the steam at different points [11].

The results of such exergy analysis have direct implication over application decision. Hence, it is necessary to identify low exergetic efficiency components of the newly installed plant. The previous attempt was made to build model using MATLAB packages which have limitation of programming skills [12]. The present paper overcomes this limitation by using excel spreadsheet approach. The exergetic efficiency of each component is determined by dividing streams into fuel and product approach. In this paper, preliminary exergy analysis is performed using an excel sheet approach to evaluate exergetic efficiency of various components such as turbines, condenser, heaters, deaerator and boiler.

## 2 Power Plant of Capacity 660 MW

The supercritical unit of capacity 660 MW is considered for exergy analysis. The pressure and temperature of steam at the inlet of the high-pressure turbine are taken as 246 bar and 565 °C, respectively. The schematic diagram of 660 MW is shown in Fig. 1. Main difference in the sub and supercritical boiler is observed in steam pressure and temperature condition at the inlet of high-pressure turbine. The once through tubes are installed instead of drum arrangement in case of the supercritical boiler. The pure sliding pressure is achieved with the help of once through boiler. The steam bled is obtained from the outlet sections of turbine to heat the feedwater. ‘B’ and ‘C’ represent the junction where all mass of steam losses get collected and

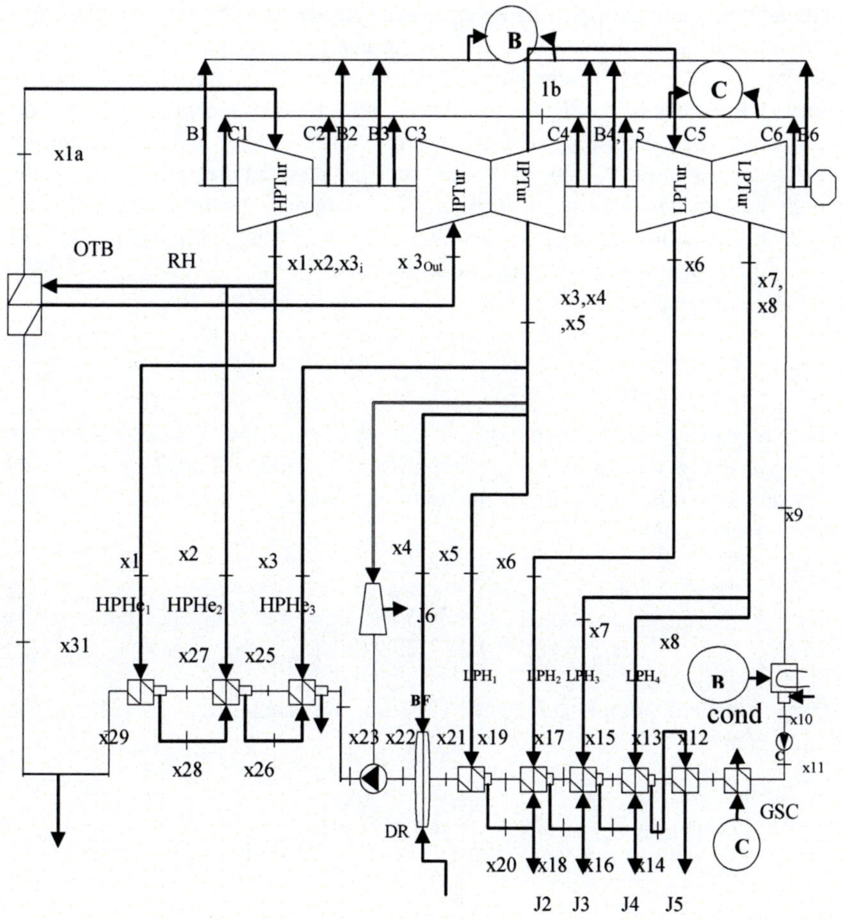


**Fig. 1** Schematic diagram of 660 MW power plant

transferred to the condenser section. The components such as condenser extraction pump, boiler feed pump, deaerator, high-pressure heaters and low-pressure heaters are incorporated to improve the performance of the overall plant.

# 3 Methodology

## *3.1 Exergy Analysis*

The exergy analysis is based upon the second law of thermodynamics and uses to identify the irreversibility in the system [13]. In the present work, the changes in the kinetic and gravitational potential energies are considered to be negligible. Certain assumptions are considered while performing energy analysis such as air is preheated to 267 °C, stack temperature of exit temp 145 °C, thermal efficiency of turbine 91%, mechanical efficiency of turbine 99.63% and generator and transformation efficiency 98.32% [14]. Inlet steams considered as fuel and outlet are considered as a product. The parameter such as temperature, pressure, specific enthalpy and specific entropy of incoming and outgoing stream is considered for evaluating energetic efficiency as given in [11, 15]. The generalized equations (Eqs. 1–5) considered for the study are as follows

$$Ex_Q + \sum_i \dot{m}_{\text{in}} \psi x_i = \sum_e \dot{m}_{\text{out}} \psi x_e + Ex_w + Ex_D \quad (1)$$

where subscripts $\psi_e$ and $\psi_i$ are the specific exergy of control volume inlet and outlet flow, respectively, referred to exergy summation of working fluid, $Ex_Q$ is the total energy due to heat involved, $Ex_w$ is the useful done on or by the system, and $Ex_D$ is the exergy destruction.

$$\psi x_i = (H_i - H_0) - 298.15 * (S_i - S_0) \quad (2)$$

$$\psi ex_e = (H_e - H_0) - 298.15 * (S_e - S_0) \quad (3)$$

Here, standard temperature and pressure are considered as 25 °C and 101.325 kPa

$$Ex_Q = \sum_{k-1}^{n} \left(1 - \frac{T_0}{T_k}\right) Q_k \quad (4)$$

where $Q_k$ is the heat transfer rate, and $T_k$ is the temperature of source or sink.

$$Ex_{X_D} = T_0 S_{\text{gen}} \quad (5)$$

where $S_{\text{gen}}$ entropy generation.

The excel spreadsheet is a powerful tool to do the complex calculation. Specific exergy is calculated for all incoming and out the coming stream from the components. Sample calculation of exergetic efficiency of high-pressure turbine in excel has been represented in [16]. The excel sheet approach was found to be a simplified approach as the operating parameters are arranged in a parallel way. Exergetic efficiency formulation of various components is given in Table 1.

**Table 1** Exergetic efficiency of various components

| Sr. no. | Name of components | Exergetic efficiency |
|---|---|---|
| 1 | High-pressure turbine | $\lambda = \frac{(m_i h_i - m_{e1} h_{e1} - m_{e2} h_{e2})}{(m_i \psi_i - m_{e1} \psi_{e1} - m_{e2} \psi_{e2})}$ |
| 2 | Re-heater | $\lambda = \frac{(m_i h_i - m_e h_e)}{(m_i \psi_i - m_e \psi_e)}$ |
| 3 | Intermediate pressure turbine | $\lambda = \frac{(m_i h_i - m_{e1} h_{e1} - m_{e2} h_{e2})}{(m_i \psi_i - m_{e1} \psi_{e1} - m_{e2} \psi_{e2})}$ |
| 4 | Low-pressure turbine | $\lambda = \frac{(m_i h_i - m_{e1} h_{e1} - m_{e2} h_{e2} - m_{e3} h_{e3} - m_{e4} h_{e4})}{(m_i \psi_i - m_{e1} \psi_{e1} - m_{e2} \psi_{e2} - m_{e3} \psi_{e3} - m_{e4} \psi_{e4})}$ |
| 5 | Condenser | $\lambda = \frac{(-m_i * T_0 * ((1-x)*(s_e - s_i) + mcw(s_{\text{in}} - s_{\text{out}}))}{((h_f - h_0) - T_0 * (s_i - s_o))}$ |
| 6 | Condenser extraction pump | $\lambda = \frac{(m_e(\psi_e - \psi_i))}{(v*(P_e - P_i)*m_i*\frac{n}{2})}$ |
| 7 | Boiler feed pump | $\lambda = \frac{(m_e(\psi_c - \psi_i))}{(P_p)}$ |
| 8 | High-pressure heater 1, 2, 3 | $\eta = 100 - \frac{(m_e \psi_e - m_i \psi_i)}{(m_{ib} \psi_{ib} - m_{eb} \psi_{eb} - \text{loss due to throttling})}$ |
| 9 | Low pressure heater 1, 2, 3 | $\eta = 100 - \frac{(m_e \psi_e - m_i \psi_i)}{(m_{ib} \psi_{ib} - m_{eb} \psi_{eb} - \text{loss due to throttling})}$ |
| 10 | Dearator | $\eta = \frac{(m_e \psi_e)}{(m_{i1} \psi_{i1} + m_{i2b} \psi_{i2} + m_{i3b} \psi_{i3})}$ |
| 11 | Boiler | $\eta = \frac{mw(\psi_e - \psi_i) + ms(\psi_{eh} - \psi_{ei})}{(m\psi)f}$ |

## 4 Result and Discussion

This paper presents the use of excel tool to evaluate the exergetic efficiency of various components. Exergy analysis is performed by categorizing incoming and outgoing stream. Condenser exergetic efficiency was found to be 34.91%, followed by the boiler and lower pressure turbine, as shown in Fig. 2. This indicates that most energy can be recovered from the condenser section.

The validation of predicted exergetic efficiency is done with the references available in the literature as shown in Table 2. The overall second law efficiency of plant is found out to be 35.28%. The tabular validation shows that predicted exergetic efficiency of components is in acceptable range with the reference values.

## 5 Conclusion

The exergy analysis of 660 MW coal-fired supercritical power plant is presented in this study. The validation of exergetic efficiency of various components with available literature has been done which is also presented in the study. The condenser contributes to lower exergetic efficiency of 34.91% followed by boiler, and low-pressure turbine has been evaluated. The low pressure in the condenser system is the reason to get low exergetic efficiency. The excel spreadsheet approach is quite simple than the other software-based approach as operating parameters are arranged

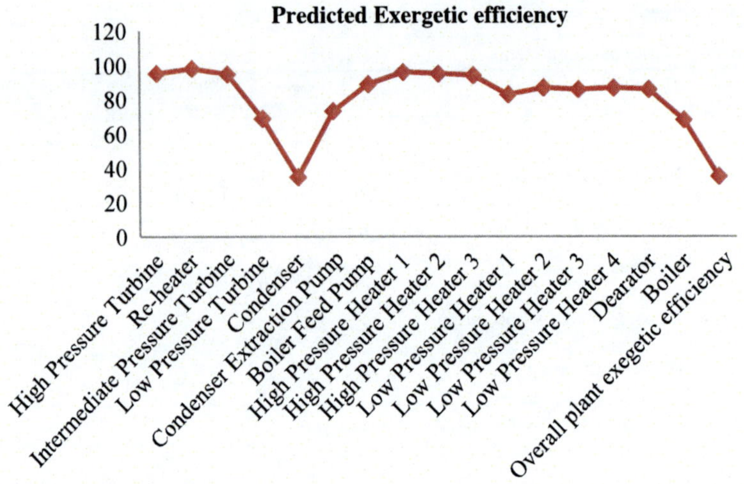


**Fig. 2** Second law efficiency of components

**Table 2** Validation of exergetic efficiency of different components

| Sr. no. | Name of components | Present study | Reference [16] | Reference [17] |
|---|---|---|---|---|
| 1 | High-pressure turbine | 95.22 | 95.04 | 96.67 |
| 2 | Intermediate pressure turbine | 94.788 | 94.8 | 95.13 |
| 3 | Low pressure turbine | 72.12 | 68.77 | 89.33 |
| 4 | Condenser | 34.91 | – | 37.93 |
| 5 | Condenser extraction pump | 86.23 | 67.18 | 95.31 |
| 6 | Boiler feed pump | 88.83 | 88.83 | 93.97 |
| 7 | High-pressure heater 1 | 95.9 | 95.9 | 98.79 |
| 8 | High-pressure heater 2 | 94.9 | 92.9 | 97.78 |
| 9 | High-pressure heater 3 | 94.01 | – | 96.51 |
| 10 | Low-pressure heater 1 | 82.8 | 89.99 | 80.85 |
| 11 | Low-pressure heater 2 | 86.7 | 86.94 | 87.63 |
| 12 | Low-pressure heater 3 | 85.71 | 86.91 | 91.29 |
| 13 | Low-pressure heater 4 | 86.6 | 85.93 | 80.85 |
| 14 | Dearator | 85.87 | – | 84.64 |
| 15 | Boiler | 68.23 | – | 62.74 |
| 16 | Overall plant exegetic efficiency | 35.28 | – | 34.38 |

in a parallel way. It is concluded that condenser is found to be unfit thermodynamic components due to low energetic efficiency. The overall exegetic efficiency of plant was calculated to 35.28%.

## References


1. Kumar R, Jilte R, Nikam K (2019) Status of carbon capture and storage in India's coal fired power plants: a critical review. Environ Technol Innov 13:94–103. https://doi.org/10.1016/j.eti.2018.10.013
2. Kumar R, Jilte R, Ahmadi MH, Kaushal R (2019) A simulation model for thermal performance prediction of a coal-fired power plant, pp 1–13. https://doi.org/10.1093/ijlct/cty059
3. Kumar R (2017) A critical review on energy, exergy, exergoeconomic and economic (4-E) analysis of thermal power plants. Eng Sci Technol Int J 20(1):283–292. https://doi.org/10.1016/j.jestch.2016.08.018
4. Kumar AKBA, Nikam KC (2020) An exergy analysis of a 250 MW thermal power plant. Renew Energy Res Appl 1(2):197–204. https://doi.org/10.22044/rera.2020.9460.1025
5. Ege A, Şahin HM (2016) Uncertainties in energy and exergy efficiency of a High Pressure Turbine in a thermal power plant. Int J Hydrogen Energy 41(17):7197–7205. https://doi.org/10.1016/j.ijhydene.2016.01.002
6. Adibhatla S, Kaushik SC (2017) Exergy and thermoeconomic analyses of 500 MWe sub critical thermal power plant with solar aided feed water heating, vol 123, pp 340–352
7. Si N et al (2017) Exergy analysis of a 1000 MW double reheat ultra-supercritical power plant. Energy Convers Manag 147:155–165. https://doi.org/10.1016/j.enconman.2017.05.045
8. Zhou J et al (2018) Exergy analysis of a 1000 MW single reheat supercritical $CO_2$ Brayton cycle coal-fired power plant. Energy Convers Manag 173(July):348–358. https://doi.org/10.1016/j.enconman.2018.07.096
9. Murav'ev VP, Kochetkov AV, Glazova EG (2016) Optimization of cooling water flow rate in nuclear and thermal power plants based on a mathematical model of cooling systems1. *Power Technol Eng* 50(3):1–4. https://doi.org/10.1007/s10749-016-0698-3
10. Dincer I, Rosen MA (2013) Exergy analysis of steam power plants. Exergy, pp 261–284. https://doi.org/10.1016/b978-0-08-097089-9.00012-7
11. Nikam KC, Kumar R, Jilte R (2020) Thermodynamic modeling and performance evaluation of a supercritical coal-fired power plant situated in Western India. Energy Sour Part A Recover Util Environ Eff 00(00):1–19. https://doi.org/10.1080/15567036.2020.1806410
12. Nikam KC, Kumar R, Jilte R (2020) Economic and exergoeconomic investigation of 660 MW coal-fired power plant. J Therm Anal Calorim. https://doi.org/10.1007/s10973-020-10213-z
13. Ameri M, Mokhtari H, Bahrami M. (2016) Energy, exergy, exergoeconomic and environmental (4E) optimization of a large steam power plant: a case study. Iran J Sci Technol—Trans Mech Eng 40(1)11–20 https://doi.org/10.1007/s40997-016-0002-z
14. Adibhatla S, Kaushik SC (2014) Energy and exergy analysis of a super critical thermal power plant at various load conditions under constant and pure sliding pressure operation. Appl Therm Eng 73(1):51–65. https://doi.org/10.1016/j.applthermaleng.2014.07.030
15. Nikam KC, Kumar R, Jilte R (2020) Exergy and exergo-environmental analysis of a 660 MW supercritical coal-fired power plant. J Therm Anal Calorim. https://doi.org/10.1007/s10973-020-10268-y
16. Kumar R, Nikam K, Jilte R (2020) A simulation model to predict coal-fired power plant production rate using artificial neural network tool. Adv Intell Syst Comput 1155:150–160. https://doi.org/10.1007/978-981-15-4029-5_15
17. Surywanshi GD, Basant Kumar Pillai B, Patnaikuni VS, Vooradi R, Anne SB (2020) Energy and exergy analyses of chemical looping combustion-based 660 MWe supercritical coal-fired power plant. Int J Exergy 31(1):14–33. https://doi.org/10.1504/IJEX.2020.104723

# A Survey Paper on Organic–Inorganic Layered Nanohybrid Embedded Polymer-based Non-volatile Resistive Switching Memory Devices

**Nipom Sekhar Das and Avijit Chowdhury**

**Abstract** Organic–inorganic and polymer-based non-volatile resistive switching memory devices have gained increasing attention because of their excellent data storage capacity, flexibility, low power consumption, durability and widely used in modern electronics. Among the materials, 2D layered nanomaterial and polymer nanocomposites have attracted increasing attention because they can easily be integrated into electronic and optoelectronic devices due to their easy fabrication process, high storage capacity and promising structural properties. Fabrication of bistable memory devices based on organic–inorganic nanohybrid with charge blocking materials and polymer nanocomposite has been broadly elucidated for their extraordinary material properties. Recent enhancement in flexible memory devices based on organic–inorganic semiconductor dispersed in ferroelectric blend polymer has been investigated thoroughly in this review paper. Recent advancement in 2D layered material, ferroelectric polymer and the development of non-volatile memristive devices is extensively explained to understand their importance in flexible devices. Outstanding memory characteristics such as current on/off ratio, $V_{set}$, $V_{reset}$ retention time, endurance of nanohybrid embedded polymer nanocomposite are also discussed meticulously. Next-generation memristor devices based on 2D nanomaterial and ferroelectric polymer with high speed and high density for developing neuromorphic computing system have been demonstrated.

**Keywords** Non-volatile memory · Organic–inorganic · Flexible · Memristive devices · 2D layered material · Ferroelectric polymer

N. S. Das · A. Chowdhury (✉)
Organic Electronics and Sensor Laboratory, Department of Physics,
National Institute of Technology Silchar, Cachar, Assam 788010, India

 
R. Kumar et al. (eds.), *Recent Trends in Thermal Engineering*, Lecture Notes in Mechanical Engineering, https://doi.org/10.1007/978-981-16-3132-0_18

# 1 Introduction

The growing need for portable and affordable non-volatile memory devices has prompted extensive research into flexible resistive switching random access memory (RRAM) devices to transcend the physical limits of conventional Si-based storage devices. Resistive random-access memory, which operates on the change of resistance states, has drawn immense attention in the area of electronic devices. Memory devices have attracted growing interest for high-density data storage, desirable flexibility, easy fabrication and low cost. Memory devices are very essential and inevitable in modern electronics due to the increasing demand of data storage function. The metal/insulator/metal structure has attracted much attention due to its potential application in resistive switching random access memory devices. Organic nanomaterial embedded polymer sandwiched between two metal electrodes has attracted promising attention in various studies of modern electronics in today's era [1–4]. RRAM memory devices have been studied most intensively in recent years, focusing on high retention, high tolerance, high speed and low power consumption [5]. The resistive switching memory properties have been largely noticed in carbon-based nanomaterials, organic materials, perovskite materials, inorganic 2D transition metal dichalcogenides (TMDs) materials, metal oxides, organic–inorganic quantum dots, hybrid nanocomposites, ferroelectric materials, etc. Carbon-based nanomaterials such as conducting graphene and its oxide show interest because of their excellent structural, physical and electrical properties. Graphene oxide shows its desirable switching characteristics in non-volatile memory devices, which is very important for the potential application in modern electronics in present era. 2D materials like graphene and $MoS_2$ are the most promising materials for electronic, optoelectronic devices due to their excellent electrical and physical properties [6]. Non-volatile resistive switching memory devices based on two-dimensional nanomaterial have emerged desirable candidate for the next-generation flexible memory devices because of their easy fabrication, high endurance cycle, long retention time and low cost. Due to excellent electrical and chemical properties of two-dimensional materials, they are broadly used in many research fields such as catalysis, energy harvesting devices and sensors. 2D inorganic semiconductors such as transition metal dichalcogenides are very promising candidate due to their layer structure, simple composition and potential advantages in many research areas including non-volatile memory devices [7]. Organic–inorganic nanohybrid embedded polymer gives a broad research in the application of non-volatile resistive memory devices. Electrical bistability is generally ascertained by the semiconductor nanomaterial. Fabricating 2D ferroelectric memory device is not easy but it can be overcome by using ferroelectric polymer combining with bendable substrates [8]. For the improvement of memory characteristics to attain low energy consumption, high speed and non-volatility, researchers are still working on it in order to store efficient amount of information in the future [9]. Moreover, 2D nanomaterials are finding broader application in memristive devices and neuromorphic systems.

## 2 Carbon-Based Nanomaterial for Non-volatile Resistive Switching Memory Devices

Carbon-based nanomaterials such as 2D graphene and graphene oxide are very promising candidate for non-volatile memory devices because of their unconventional structural and electrical properties. Graphene has drawn increasing attention in the field of material science due to its impressive chemical, structural and mechanical properties. Graphene is a 2D layer honeycomb-like structure having $sp^2$ hybridized carbon atoms [10]. Graphene oxide which is electrically insulating prepared by the oxidation of graphene sheets. By controlling the functional groups, electronic state can be easily tuned. Due to the presence of oxygen-containing functional groups like hydroxyl, carboxylic, epoxy, carbonyl groups make GO insulating in nature and suitable for the potential application in non-volatile memory devices. Excellent transparency, flexibility, easy fabrication process and electronic property make GO a promising candidate in modern electronic, transistors and optoelectronic devices [11]. Graphene oxide combined with polymer matrix enhances the memory properties like low voltage, high endurance cycle, long retention time, and the memory characteristics of different memory devices are shown in Table 1 (Figs. 1, 2).

## 3 2D Transition Metal Dichalcogenides Embedded Ferroelectric Blend Polymer-Based Nanocomposite for Non-volatile Resistive Memory Devices

Polymer nanocomposite is a combination of a polymer matrix and inclusions that have at least one dimension in nanometer size range known as polymer nanocomposite. Some advantages of polymer nanocomposites are superior mechanical properties, structural and thermal stability, promising electrical conductivity. Polymers are commonly invigorated by various sized fillers to abate some of the limitations of polymers and hence to widen their applications [19]. Some examples of polymers used in resistive random access memory are polystyrene (PS), poly(methylmethacrylate) (PMMA), polyvinylpyrrolidone (PVP), poly (N-vinylcarbazole) (PVK), polyimide (PI), polyaniline(PANI), triphenylamine-poly (azomethine) (TPAPAM), ferroelectric polymer such as poly(vinylidene fluoride) (PVDF) and poly(vinylidene fluoride-co-trifluoroethylene) P(VDF-TrFE). Polymer nanocomposites are broadly applicable in many fields, for example, sensors, LEDs, photodiodes, memory devices, etc., [20–22].

**Table 1** Non-volatile memory devices using carbon-based nanomaterial and polymer nanocomposite

| Authors | Device structure | Fabrication process | $I_{ON}/I_{OFF}$ | $V_{set}$ (volt) | $V_{reset}$ (volt) | Retention (sec) | Endurance cycles | Explanation | References | Publication year |
|---|---|---|---|---|---|---|---|---|---|---|
| Khurana et al. | Pt/GO/ITO | Spin coating | $\sim 10^4$ | −1.2 to −1.8 | 3–3.4 | $10^4$ | $10^2$ | The device showed stable bipolar resistive switching behavior composed of graphene oxide thin film on ITO-coated substrate with good endurance and retention characteristics | [12] | 2013 |
| Khurana et al. | ITO/GO-ZnO/Al | Spin coating | $10^2$ | 2.1 | −2.0 | $10^4$ | $10^2$ | The device showed bipolar resistive switching with lower switching voltages and maintained constant resistance ratio. The device showed good retention and endurance | [13] | 2014 |
| Jeong et al. | Al/GO/Al | Spin casting | $>10^2$ | −2.5 | 2.5 | $10^5$ | $10^2$ | Graphene oxide thin film was prepared by spin casting, and the device showed the bipolar resistive switching characteristics with high retention time ($10^5$ s.) | [14] | 2010 |
| He et al. | Cu/GO/Pt | Vacuum filtration method | 20 | 0.8 | −0.75 | $>10^4$ | $10^2$ | Graphene oxide thin film was prepared on Pt substrate, and the device showed the reversible bipolar resistive switching behavior with excellent retention and low switching voltages | [15] | 2009 |

(continued)

**Table 1** (continued)

| Authors | Device structure | Fabrication process | $I_{ON}/I_{OFF}$ | $V_{set}$ (volt) | $V_{reset}$ (volt) | Retention (sec) | Endurance cycles | Explanation | References | Publication year |
|---|---|---|---|---|---|---|---|---|---|---|
| Wu et al. | Ag/PI/GO: PI/PI/ITO | Spin coating | $10^5$ | 3.5 | – | $10^3$ | $10^2$ | The device showed the multilevel resistive switching properties. The device exhibited high current on/off ratio and good endurance. From the results mentioned in the table, it can be concluded that GO-PI hybrid composite has attracted promising application in non-volatile memory devices | [16] | 2011 |
| Li et al. | ITO/PEDOT + GO/Al | Spin coating | $10^4$ | 1.5 | −3.6 | $10^4$ | – | The device exhibited the reversible bistable resistance switching with high on/off current ratio and good retention time | [17] | 2016 |
| Gogoi et al. | ITO/rGO-CdS/PMMA/Al | Spin coating | $10^6$ | −0.48 | – | $10^4$ | $10^3$ | The device showed bipolar resistive switching characteristics with high on/off current ratio ($10^6$), low switching voltage and very low set/reset power density. High retention time and endurance which confirms the stability and repeatability of the device | [18] | 2020 |

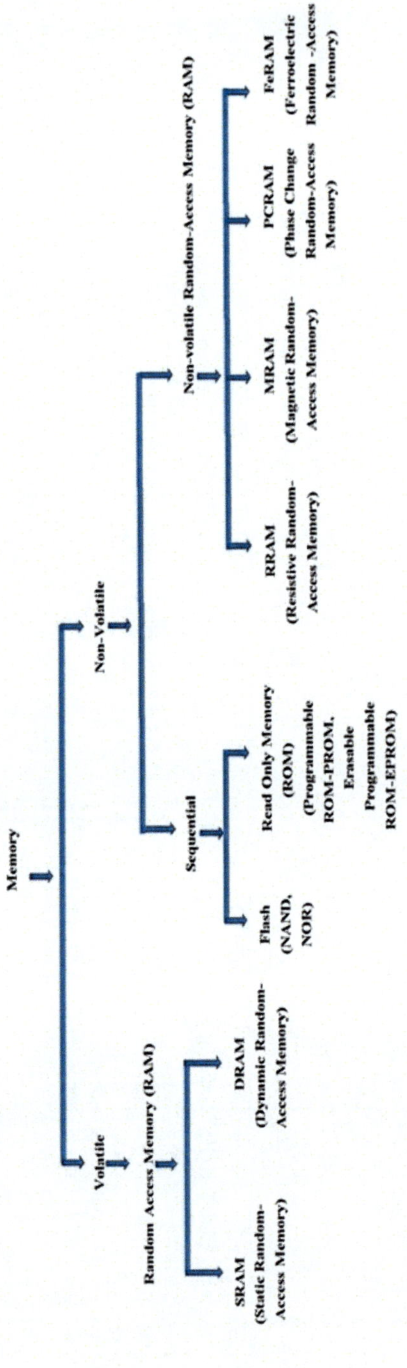


**Fig. 1** Classification of memory

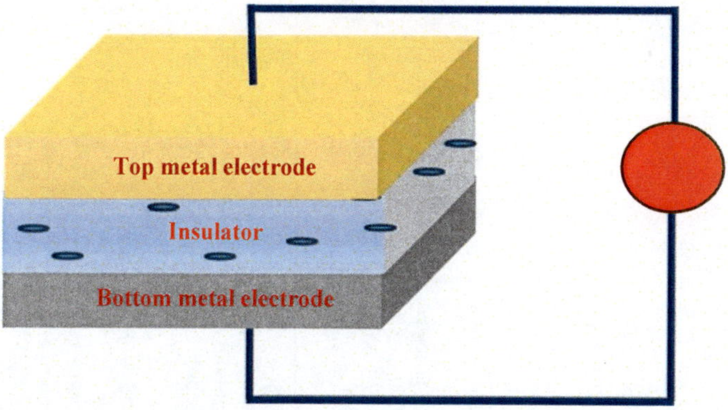


**Fig. 2** Resistive switching random access memory device structure

### 3.1 Ferroelectric Polymer Material

The most known ferroelectric polymer is poly (vinylidene fluoride) (PVDF), which is a semi-crystalline thermoplastic fluoropolymer with monomer ($CH_2$-$CF_2$), where two hydrogen and two fluorine atoms are bonded to carbon atom. It shows ferroelectric, piezoelectric and pyroelectric properties. There are many crystallization phases present such as orthorhombic $\alpha$, $\beta$, $\gamma$ phases and a monoclinic $\delta$ phase. Recently, ferroelectric polymers have attracted growing interest in modern electronics, energy harvesting devices and sensors due to its promising material behavior. To enrich the ferroelectric properties of PVDF, TrFE and CTFE, polymers are used as co-polymers [23, 24]. Ferroelectric polymeric materials have been regarded as the most promising candidate for developing memristor devices for information storage and constructing artificial synapse in neuromorphic computing applications (Table 2).

### 3.2 2D Transition Metal Dichalcogenides Material (TMDs)

2D nanosheets transition metal dichalcogenides (TMDs) materials, for example, $MoS_2$, $WS_2$, $MoSe_2$, $WSe_2$, etc. have gained increasing interest because of their potential applications in electronic, sensors and optoelectronic devices [29, 30]. They have been regarded as the growing candidate because of their excellent physical, electronic and chemical properties. Recent progresses of 2D layer materials in physics and technology have been extensively explored because of their promising performances in modern electronic such as transistors and resistive memory devices [31]. From last decade, it has been observed that making nanohybrid material with 2D graphene sheets such as graphene oxide (GO) and reduced graphene oxide (rGO) enhance the chemical, physical and electronic properties which lead nanohybrid material an excellent candidate in many applications (Table 3).

**Table 2** Non-volatile resistive switching memory devices based on ferroelectric blend polymer

| Authors | Device structure | Fabrication process | $I_{ON}/I_{OFF}$ | $V_{set}$ (volt) | $V_{reset}$ (volt) | Retention (sec) | Endurance cycles | Explanation | References | Publication Year |
|---|---|---|---|---|---|---|---|---|---|---|
| Hyun et al. | Au/P (VDF-TrFE)/ ITO | Spin coating | – | 5.8 | −6.1 | – | $10^3$ | The device showed the diode behavior and ferroelectric piezoresponse hysteresis loop. The diode property was maintained after the large number of bendings | [25] | 2016 |
| Xia et al. | Al/ P (VDF-TrFE)/ PTFE/Al | Spin coating | $10^5$ | 3.4 | −2.6 | $10^3$ | – | Here, high-quality fabrication of ferroelectric polymer thin film has been reported. Organic semiconducting/ferroelectric blend films have been fabricated via removable PTFE template method which is expected to be used for other organic semiconductors | [26] | 2017 |
| Kim et al. | Au/ P (VDF-TrFE)/ ZnO/Si | Spin coating | $10^7$ | −3.9 to −2.74 | 2.81–2.60 | $10^4$ | $10^4$ | Here, organic ReRAM showed bipolar resistive switching with a very high on/off current ratio and also showed good retention and endurance | [27] | 2018 |
| Wang et al. | Ag/PVDF/ PMMA/ F8T2/Ag | Spin coating | $10^3$ | – | – | $10^3$ | – | In this study, the ternary film exhibited good resistive switching performance with good retention. Here, it is seen that when PMMA doped with PVDF, then ferroelectric phase transformation occurs | [28] | 2018 |

**Table 3** Non-volatile resistive switching memory devices based on two-dimensional transition metal dichalcogenides embedded ferroelectric blend polymer

| Authors | Device structure | Fabrication process | $I_{ON}$/ $I_{OFF}$ | $V_{set}$ (volt) | $V_{reset}$ (volt) | Retention (sec) | Endurance cycles | Explanation | References | Publication year |
|---|---|---|---|---|---|---|---|---|---|---|
| Rehman et al. | Ag/ $MoS_2$-PVA/ Ag | EHDA | $10^2$ | 3 | −3 | $10^5$ | $>10^3$ | In this study, the fabricated device exhibited the bistable, non-volatile and rewritable resistive switching behavior at low operating voltage and showed high retention and good endurance | [32] | 2016 |
| Rehman et al. | Ag/ $WS_2$/Ag | EHDA | $10^3$ | 2 | −2 | $10^5$ | 1500 | Here, the device exhibited the non-volatile bipolar behavior and showed a long retention time | [33] | 2017 |
| Lee et al. | Al/$WS_2$ NSs: PMMA/ ITO/ PEN | Spin coating | $10^4$ | 0.4–0.6 | −4.3 to −4.5 | $10^4$ | $>10^2$ | The memristive device showed the bipolar switching behavior with low power consumption and also exhibited the decent retention time | [34] | 2019 |

## 4 Conclusions and Perspectives

From the review, it can be concluded that organic–inorganic nanohybrid embedded polymer nanocomposite acts as a pivotal role in many research areas because of their easy fabrication, low cost and layer structure. Briefly, the recent advances in memory devices based on 2D carbon nanomaterial, transition metal dichalcogenides and polymer nanocomposite have been reviewed, and attempts have been made to find a broader perspective on them in modern electronic devices. Unique chemical, physical and electronic properties of two-dimensional materials and ferroelectric polymers have gained great research interest and become the building block for many physical and technological applications such as electronic, optoelectronic, energy harvesting devices and sensor technology. From the survey of the paper, it can be concluded that high endurance and retention properties of the devices are suitable for future non-volatile memory applications. These results indicate that high on/off current ratio, high retention and lower switching voltage can find promising applications in next-generation high-density multilevel non-volatile resistive switching and also can construct artificial synapse based on memristive devices for neuromorphic computing.

## References

1. Khurana G, Misra P, Katiyar RS (2014) Multilevel resistive memory switching in graphene sandwiched organic polymer heterostructure. Carbon 76:341–347
2. Gogoi KK, Das NS, Chowdhury A (2019) Tuning of electrical hysteresis in PMMA/GOs/PMMA multi-stacked devices. Mater Res Express 6:085108
3. Gogoi KK, Chowdhury A (2019) Performance enhancement of solution processed organic memories by exploiting synergistic organic-inorganic hybrid composites. J Phys Chem C 124 (1):1108–1120
4. Gogoi KK, Das R, Paul T, Ghosh S, Chowdhury A (2019) Tuning of bipolar resistive switching and memory characteristics of cadmium sulphide nanorods embedded in PMMA matrix. Mater Res Express 6:115107
5. Sun Y, Wen D, Bai X, Lu J, Ai C (2017) Ternary resistance switching memory behavior based on graphene oxide embedded in a polystyrene polymer layer. Sci Rep 7(1):3938
6. Tan C, Liu Z, Huang W, Zhang H (2015) Non-volatile resistive memory devices based on solution-processed ultrathin two-dimensional nanomaterials. Chem Soc Rev 44:2615–2628
7. Liu J, Zeng Z, Cao X, Lu G, Wang L-H, Fan Q-L, Huang W, Zhang H (2012) Preparation of $MoS_2$-polyvinylpyrrolidone nanocomposites for flexible non-volatile rewritable memory devices with reduced graphene oxide electrodes. Small 8(22):3517–3522
8. Li H, Wang R, Han S-T, Zhou Y (2020) Ferroelectric polymers for non-volatile memory devices: a review. Polym Int 69:533–544
9. Ma C, Luo Z, Huang W, Zhao L, Chen Q, Lin Y, Liu X, Chen Z, Liu C, Sun H, Jin X, Yin Y, Li X (2020) Sub-nanosecond memristor based on ferroelectrictunnel junction. Nat Commun 11(1):1439
10. Allen MJ, Tung VC, Kaner RB (2010) Honeycomb carbon: a review of graphene. Chem Rev 110:132–145

11. Chen Y, Zhang B, Liu G, Zhuang X, Kang E-T (2012) Graphene and its derivatives: switching ON and OFF. Chem Soc Rev 41:4688–4707
12. Khurana G, Misra P, Katiyar RS (2013) Forming free resistive switching in graphene oxide thin film for thermally stable nonvolatile memory applications. J Appl Phys 114:124508
13. Khurana G, Misra P, Kumar N, Katiyar RS (2014) Tunable power switching in non-volatile flexible memory devices based on graphene oxide embedded with ZnO nanorods. J Phys Chem C 118:21357–21364
14. Jeong HY, Kim JY, Kim JW, Hwang JO, Kim J-E, Lee JY, Yoon TH, Cho BJ, Kim SO, Ruoff RS, Choi S-Y (2010) Graphene oxide thin films for flexible nonvolatilememory applications. Nano Lett 10:4381–4386
15. He CL, Zhuge F, Zhou XF, Li M, Zhou GC, Liu YW, Wang JZ, Chen B, Su WJ, Liu ZP, Wu YH, Cui P, Li R-W (2009) Nonvolatile resistive switching in graphene oxide thin films. Appl Phys Lett 95:232101
16. Wu C, Li F, Zhang Y, Guo T, Chen T (2011) Highly reproducible memory effect of organic multilevel resistive-switch device utilizing graphene oxide sheets/polyimide hybrid nanocomposite. Appl Phys Lett 99:042108
17. Li Y, Ni X (2016) One-step preparation of graphene oxide-poly(3,4ethylenedioxythiophene) composite films for nonvolatile rewritable memory. RSC Adv 6:16340–16347
18. Gogoi KK, Chowdhury A (2020) Performance improvement of organic resistive memories by exploiting synergistic layered nanohybrid dispersed polymercomposites. J Appl Phys 127:065501
19. Fu S, Sun Z, Huang P, Li Y, Hu N (2019) Some basic aspects of polymer nanocomposites: a critical review. Nano Mater Sci 1:2–30
20. Gogoi KK, Chowdhury A (2019) Electric field induced tunable electrical hysteresis in poly (methyl methacrylate)/graphene oxide heterostructures. In: AIP Conference Proceedings, vol 2100, pp 020022
21. Liu G, Zhuang X, Chen Y, Zhang B, Zhu J, Zhu C-X, Neoh K-G, Kang E-T (2009) Bistable electrical switching and electronic memory effect in a solution-processable graphene oxide-donor polymer complex. Appl Phys Lett 95:253301
22. Wu C, Li F, Guo T (2014) Efficient tristable resistive memory based on single layer graphene/ insulating polymer multi-stacking layer. Appl Phys Lett 104:183105
23. Lenz T, Dehsari HS, Asadi K, Blom PWM, Groen WA, Leeuw DM (2016) Thin film thermistor with positive temperature coefficient of resistance based on phase separated blends of ferroelectric and semiconducting polymers. Appl Phys Lett 109:133302
24. Xie B, Zhang Q, Zhang L, Zhu Y, Guo X, Fan P, Zhang H (2018) Ultrahigh discharged energy density in polymernanocomposite by designing linear/ferroelectricbilayer heterostructure. Nano Energy 54:437–446
25. Hyun S, Kwon O, Choi C, Joseph KLV, Kim Y, Kim JK (2016) Self-positioned nanosized mask for transparent and flexible ferroelectric polymer nanodiodes array. ACS Appl Mater Interfaces 8:27074–27080
26. Xia W, Peter C, Weng J, Zhang J, Kliem H, Jiang Y, Zhu G (2017) Epitaxy of ferroelectric P (VDF-TrFE) films via removable PTFE templates and its application in semiconducting/ ferroelectric blend resistive memory. ACS Appl Mater Interfaces 9(13):12130–12137
27. Kim TY, Anoop G, Son YJ, Kim SH, Lee E, Jo JY (2018) Ferroelectric-mediated filamentary resistive switching in P(VDFTrFE)/ZnO nanocomposite films. Phys Chem Chem Phys 00:1–8
28. Wang H, Wang C-T, Xu F, Yang J, Liu J, Cai W, Zhu G (2018) Resistive switching and nanoscale chemical mapping of phase separation in PVDF/PMMA/F8T2 ternary thin films. Polymer 153:498–506
29. Yin Z, Zeng Z, Liu J, He Q, Chen P, Zhang H (2012) Memory devices using a mixture of $MoS_2$ and graphene oxide as the active layer. Small 9(5):721–731
30. Ye L, Xu H, Zhang D, Chen S (2014) Synthesis of bilayer $MoS_2$ nanosheets by a facile hydrothermal method and their methyl orange adsorption capacity. Mater Res Bull 55:221–228

31. Tan C, Zhang H (2015) Two-dimensional transition metal dichalcogenide nanosheet-based composites. Chem Soc Rev 44(9):2713–2731
32. Rehman MM, Siddiqui GU, Gul JZ, Kim S-W, Lim JH, Choi KH (2016) Resistive switching in all-printed, flexible and hybrid $MoS_2$-PVA nanocomposite based memristive device fabricated by reverse offset. Sci Rep 6(1):36195
33. Rehman MM, Siddiqui GU, Doh YH, Choi KH (2017) Highly flexible and electroforming free resistive switching behavior of tungsten disulfide flakes fabricated through advanced printing technology. Semicond Sci Technol 32:095001
34. Lee JH, Wu C, Sung S, An H, Kim TW (2019) Highly flexible and stable resistive switching devices based on $WS_2$ nanosheets: poly(methylmethacrylate) nanocomposites. Sci Rep 9(1): 19316

# Thermo-Hydraulic CFD Analysis of Impinging Jet Solar Air Heater with Different Jet Geometries

**Siddhita Yadav and R. P. Saini**

**Abstract** Renewable energy is an environment friendly alternative to fossil fuels. Sun is the ultimate source of energy, and with the help of solar air heater, solar energy can be directly utilized for different solar thermal applications. However, it needs performance enhancement for better efficiency. Present study shows the thermal behavior of impinging jet solar air heater with five different jet geometries such as triangular, rectangular, square, hexagonal and circular jet. Heat transfer as Nusselt number and pumping loss as friction factor have been analyzed and compared for all the geometries under similar operating conditions. ANSYS FLUENT 18.1 is used to perform computational analysis. RNG $k$-$\varepsilon$ turbulence model is used to solve transport equations. A comparative study has been performed for different jet geometries under a constant heat flux of 1000 W/m$^2$ and Reynolds number ranging from 3700 to 16,500. Results show the maximum and minimum enhancement in heat transfer of 1.66 and 1.14 times for circular and triangular shape impinging solar air heater, respectively. The maximum value of the thermo-hydraulic performance parameter is obtained as 1.16 for circular jet impinging solar air heater at Reynolds number 16500.

**Keywords** Heat transfer · CFD · Turbulence model · Jet impingement

## 1 Introduction

Fossil fuel is depleting with very fast rate, and it takes million years to replenish these energy resources. Thus, energy conservation is the primary concern for today's fast-moving world. Non-renewable energy resources are the best alternative to reserve the depleting fossil fuels. Solar energy is one of the best energy sources

S. Yadav (✉) · R. P. Saini
Department of Hydro and Renewable Energy, IIT Roorkee, Uttarakhand 247667, India
e-mail: syadav@ah.iitr.ac.in

R. P. Saini
e-mail: rajsafah@iitr.ac.in



R. Kumar et al. (eds.), *Recent Trends in Thermal Engineering*, Lecture Notes in Mechanical Engineering, https://doi.org/10.1007/978-981-16-3132-0_19

as it is free of cost, environmental friendly source. Solar energy can be trapped and converted into thermal energy utilizing solar collectors. Solar air heater (SAH) is one of the simple and economic application of solar collectors which use environmental air as working fluid [1]. It is playing an important role in different applications like buildings heating, grain drying. With advantages of cheap and easy construction, no freezing and corrosion problems, SAH has disadvantage of low heat transfer capability. Heat transfer in SAH can be improved by modifying absorber design like roughness on absorber, providing jet plate, storage material inside the duct [2].

Jet impingement techniques are introduced by Kercher and Tabakoff [3] using round shape jets in a square array. Further, Metzger et al. [4] investigated that the heat transfer characteristics of inline jets are better than the staggered jet pattern. Choudhury and Garg [5] applied impinging jet technique on solar air heaters experimentally. Later on, Xing et al. [6] found that the heat transfer in inline circular jet is better than staggered jet array. Further, Chauhan and Thakur [7, 8] presented the thermo-hydraulic performance of impinging jet solar air heater (SAHJI). Further, Chauhan et al. [9, 10] optimized parameters of SAHJI using PSI and Taguchi method. Matheswaran et al. [11] proposed energy and exergy analysis of SAHJI, and recently, Shetty et al. [12] investigated perforated absorber type SAH experimentally. Gawande et al. [13] and Singh and Singh [14] investigated different SAH configuration using CFD tool but none of them performed CFD analysis specifically on SAHJI.

Based on the literature review, it is found that previous studies are focused on the experimental investigations, and very few researchers have utilized CFD tool to study the solar air heaters but CFD can be a useful tool for cost cutting. Keeping this in view, under the present study, a CFD analysis is performed on SAHJI with five different jet geometry (triangular, rectangular, square, hexagonal and circular) using ANSYS-18.1. Results from CFD analysis are validated by previously available equations in the literature. Heat transfer, frictional losses and thermo-hydraulic performance of SAH with different jet geometries are compared under similar operating conditions. This study can be useful for further researches on SAHJI.

## 2 Methodology

In the present study, a duct with 1 m length and $0.3 \times 0.025$ m$^2$ cross section has been modeled with entry and exit length as per ASHRAE guidelines [15]. The duct consists of a smooth absorber plate with a uniform heat flux of 1000 W/m$^2$ and a jet plate with inline jet array. Bottom and sides are properly insulated as shown in Fig. 1.

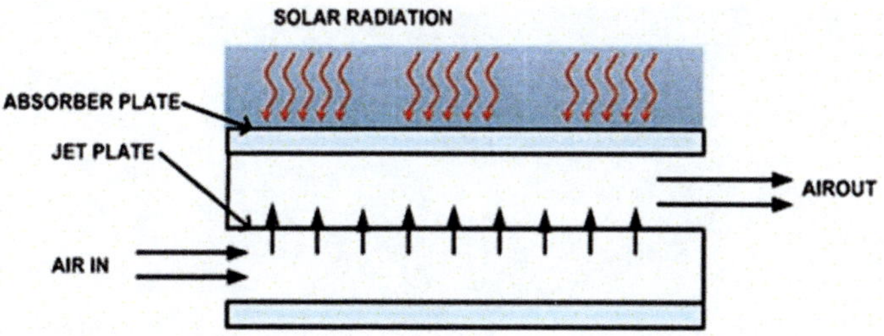


**Fig. 1** Schematic of SAHJI

## 2.1 Model Geometry

3D model of SAHJI has been generated on ANSYS workbench. Five types of jet plates configuration with different jet shapes (triangular (TSAHJI), rectangular (RSAHJI), square (SSAHJI), hexagonal (HSAHJI) and circular (CSAHJI)) have been considered for the present study out of which all the sides are equal to 3 mm for triangular, rectangular, square, hexagonal shape jets, whereas for circular jets, diameter of jet is considered as 3 mm. Spanwise and streamwise pitch is 0.866 and 1.732, respectively, for each jet plate as shown in Fig. 2.

## 2.2 Model Meshing

Fine, unstructured and non-uniform grid mesh are generated using ANSYS-ICEM. To decide the optimum grid size, grid independence test is performed. Best suitable grid size for TSAHJI, RSAHJI, SSAHJI, HSAHJI and CSAHJI is found as 2.12, 2.08, 2.09, 1.87, 1.81 million for present model because beyond this value, the uncertainty in results (Nu) is less than 1%. Detailed GIT for CSAHJI is given in Table 1. To visualize the proper flow at absorber at inlet and outlet, 12 inflation layers and 1.01 of growth rate are provided. Flow domain mesh for CSAHJI is presented in Fig. 3.

## 2.3 Applied Model and Boundary Conditions

As per the problem, boundary conditions were provided using FLUENT solver. Rectangular duct consists of an entry, exit and test sections. Test section is considered as absorber plate having a uniformly distributed heat flux of 1000 $W/m^2$ with no slip boundary condition. Velocity inlet and pressure outlet boundary conditions are taken at entry and exit of air respectively. For side and bottom walls, adiabatic-no slip boundary condition has been used. The simulation was carried out at different values of Reynolds number ranging 3700–16,500 (6 steps). In order to

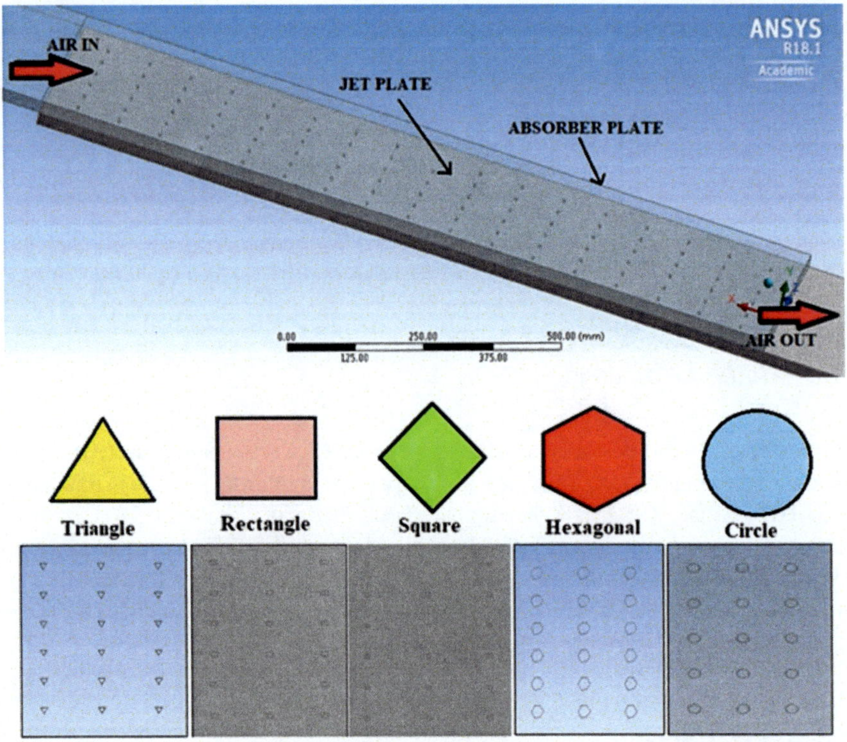


**Fig. 2** 3D modeling of impinging jet solar air heater with different jet geometries

**Table 1** Grid independence test for CSAHJI for Re = 3500

| No. of element | Nu | % change in Nu |
|---|---|---|
| 0.95 | 13.37 | 23.095 |
| 1.43 | 17.384 | 18.154 |
| 1.81 | 21.24 | 0.7012 |
| 1.94 | 21.39 | 0.604 |
| 2.14 | 21.52 | – |

solve the governing equations, RNG k-ε turbulence model was used in simulation. A convergence criteria of $1 \times 10^{-4}$ was provided for the residuals. The results are obtained using CFD POST.

## 3 Results and Discussion

CFD model of smooth solar air heater (SSAH) has been validated with results obtained by the previously provided equations in the literature under similar operating conditions. Thermal behavior of different SAHJI configurations has been

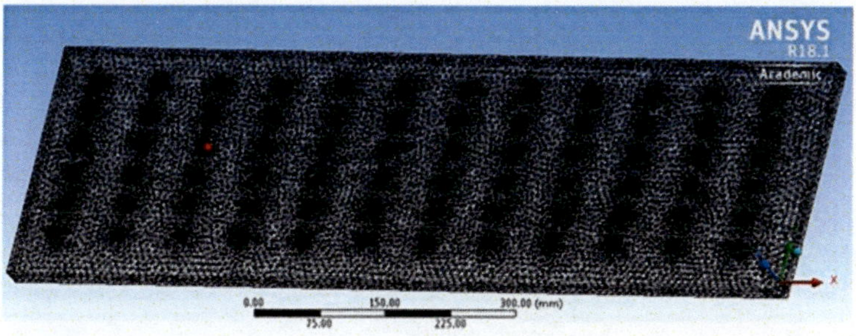


**Fig. 3** Meshing of CSAHJI

compared with SSAH, and results obtained from the comparison have been discussed. Variation of temperature with increasing Re values has also been presented using contours.

## 3.1 *Validation*

The average absolute deviation of Nusselt number between CFD results and Dittus-Boelter equation [16] is found as 4.67%. The reason of this deviation of CFD results with equations result is may be due to assumptions and round-off error in higher-order equations. CFD results show acceptable agreement with results obtained from previously done research data as shown in Fig. 4.

## 3.2 *Contours for Temperature Distribution*

Figure 5 shows the contours of temperature for smooth and impinging jet duct, respectively, for Reynolds number of 3700 and 16,500. It is observed that with the increase in Reynolds number, temperature of absorber plate is decreased while

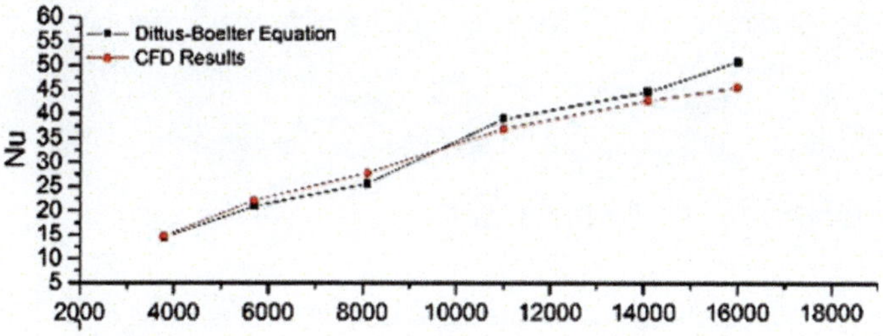


**Fig. 4** Validation of CFD results

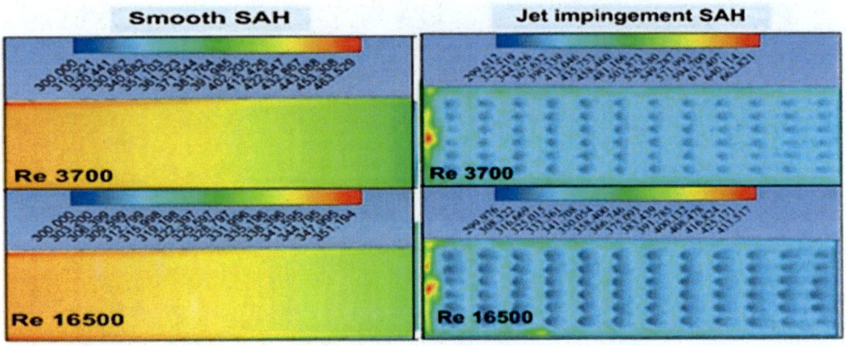


**Fig. 5** Contour for temperature (K) for SSAH, CSAHJI at Re = 3700 and 16,500

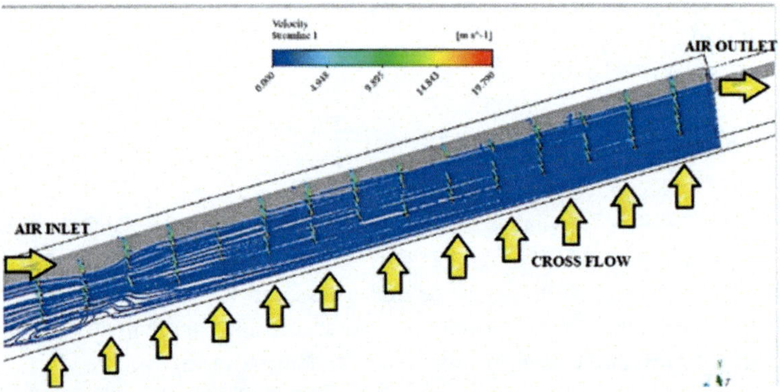


**Fig. 6** Streamlines inside the duct for SAHJI

pressure inside the duct is increased. Due to the generation of laminar sub-layer over the absorber surface, the temperature of plate inside the duct increases throughout the duct length.

For SAHJI, temperature is higher at the corners of the plate and minimum, where the jet strikes on the absorber plate. The cross-flow of the jet on absorber plate breaks the laminar sub-layer, and heat transfer is increased between absorber and fluid flowing inside the duct. Thus, it is found that heat transfer rate is higher for SAHJI than smooth SAH.

## *3.3 Streamlines Inside the Duct*

Figure 6 shows streamlines inside the SAHJI duct. Cross-flow of air through jet plate is encouraging the heat transfer inside the duct, and air is absorbing more heat

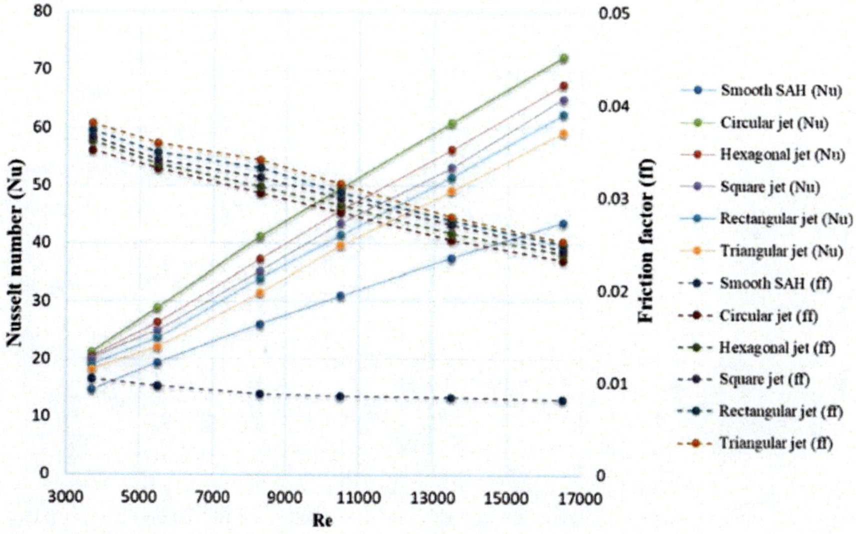


**Fig. 7** Reynolds number variation with Nusselt number (Nu) and friction factor (ff)

with heated surface due to jet impact on heated surface. Jet strike on absorber is breaking laminar sub-layer and increasing turbulence inside the duct. With the increase in Reynolds, the interaction time of heated surface and air is reduced. Thus, heat transfer is decreased at a very high velocity of inlet air. Air absorbs highest heat in case of CSAHJI and minimum for TSAHJI under similar conditions.

## *3.4 Thermo-Hydraulic Performance*

Nu increases with an increase in Re while friction factor decreases as shown in Fig. 7. The maximum range of Nu is noticed for CSAHJI, while maximum ff is found for TSAHJI for the considered range of Re. With the increase in heat transfer, the pumping power requirement is also increased. Figure 8 shows the maximum and minimum value of enhancement in heat transfer (NNE) which is found as 1.66 at Re = 16,500 and 1.14 at Re = 5500 for CSAHJI and TSAHJI, respectively, and corresponding friction factor increment (FFI) is found as 2.86 and 3.74. The maximum and minimum value of thermo-hydraulic performance parameter [THPP = {(NNE)/ (FFI)^$^{1/3}$}] has been observed as 1.16 and 0.73 for CSAHJI and TSAHJI. The value of THPP is more than 1.0 for Re ranging from 8300 to 16,500 for CSAHJI, which indicates friction factor and is dominated by heat transfer for this range of parameters. Thus, the order of overall heat transfer enhancement in different configuration of SAHJI is as follows:

CSAHJI< RSAHJI < SSAHJI < HSAHJI < TSAHJI.

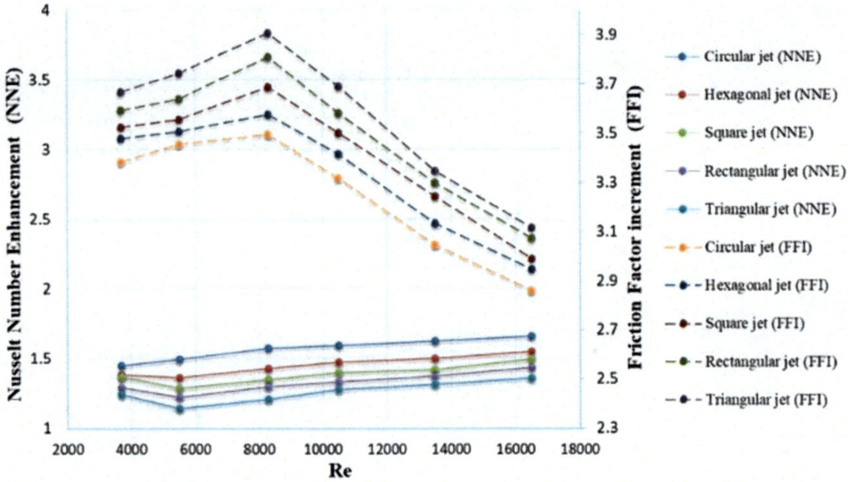


**Fig. 8** Reynolds number variation with Nusselt number enhancement (NNE) and friction factor increment (FFI)

# 4 Conclusions

In the present study, a computational CFD analysis has been made to study the thermo-hydraulic performance of SAHJI with different jet configuration using ANSYS-18.1. In the present study, CFD results show good agreement with previously provided mathematical equations for the SSAH. Based on the comparison, CFD tool is found to be a reliable technique to study the solar air heater.

From the simulation results, it is found that enhancement in Nusselt number increases with Reynolds number from 1.43 to 1.66 for CSAHJI and from 1.24 to 1.35 times for TSAHJI in the range of parameters considered. The increasing order of heat transfer is found as TSAHJI < RSAHJI < SSAHJI < HSAHJI < CSAHJI. Thus, CSAHJI is found to be more efficient in comparison with other configuration in terms of heat transfer.

Friction factor decreases as the Reynolds number increases. The increment in friction factor is observed from 3.38 to 2.86 and 3.66 to 3.11 for CSAHJI and TSAHJI, respectively. The power required to propel the air inside the duct is the function of friction factor, and it is observed as CSAHJI < HSAHJI < SSAHJI < RSAHJI < TSAHJI.

Further, maximum value of THPP has been found as 1.16 for CSAHJI correspond to Reynolds number of 16,500 in the range of parameters considered under present study. Heat transfer is found dominating over the friction factor inside the duct as THPP is greater than unity. Overall, heat transfer behavior of circular jet solar air heater with jet impingement (CSAHJI) is found to be better than other four configuration and SSAH with low pumping power.

## References


1. Razak AA et al (2016) Review on matrix thermal absorber designs for solar air collector. Renew Sustain Energy Rev 64:682–693. https://doi.org/10.1016/j.rser.2016.06.015
2. Prakash C, Saini RP (2019) Heat transfer and friction in rectangular solar air heater duct having spherical and inclined rib protrusions as roughness on absorber plate. Exp Heat Transf 32(5):469–487. https://doi.org/10.1080/08916152.2018.1543367
3. Kercher DM, Tabakoff W (1969) Heat transfer by a square array of round air jets impinging perpendicular to a flat surface including the effect of spent air. ASME-Paper 69-GT-4
4. Metzger DE, Florschuetz LW, Takeuchi DI, Behee RD, Berry RA (1979) Heat transfer characteristics for inline and staggered arrays of circular jets with crossflow of spent air. J Heat Transf 101(3):526–531. https://doi.org/10.1115/1.3451022
5. Choudhury C, Garg HP (1991) Evaluation of a jet plate solar air heater. Sol Energy 46 (4):199–209. https://doi.org/10.1016/0038-092X(91)90064-4
6. Xing Y, Spring S, Weigand B (2010) Experimental and numerical investigation of heat transfer characteristics of inline and staggered arrays of impinging jets. J Heat Transf 132 (9):1–11. https://doi.org/10.1115/1.4001633
7. Chauhan R, Thakur NS (2013) Heat transfer and friction factor correlations for impinging jet solar air heater. Exp Therm Fluid Sci 44:760–767. https://doi.org/10.1016/j.expthermflusci.2012.09.019
8. Chauhan R, Thakur NS (2014) Investigation of the thermohydraulic performance of impinging jet solar air heater. Energy 68:255–261. https://doi.org/10.1016/j.energy.2014.02.059
9. Chauhan R, Singh T, Thakur NS, Patnaik A (2016) Optimization of parameters in solar thermal collector provided with impinging air jets based upon preference selection index method. Renew Energy 99:118–126. https://doi.org/10.1016/j.renene.2016.06.046
10. Chauhan R, Singh T, Kumar N, Patnaik A, Thakur NS (2017) Experimental investigation and optimization of impinging jet solar thermal collector by Taguchi method. Appl Therm Eng. https://doi.org/10.1016/j.applthermaleng.2017.01.025
11. Matheswaran MM, Arjunan TV, Somasundaram D (2018) Analytical investigation of solar air heater with jet impingement using energy and exergy analysis. Sol Energy. https://doi.org/10.1016/j.solener.2017.12.036
12. Shetty SP, Paineni A, Kande M, Madhwesh N, Sharma NY, Karanth KV (2020) Experimental investigations on a cross flow solar air heater having perforated circular absorber plate for thermal performance augmentation. Sol Energy 197(January):254–265. https://doi.org/10.1016/j.solener.2020.01.005
13. Gawande VB, Dhoble AS, Zodpe DB, Chamoli S (2016) Experimental and CFD investigation of convection heat transfer in solar air heater with reverse L-shaped ribs. Sol Energy 131:275–295. https://doi.org/10.1016/j.solener.2016.02.040
14. Singh I, Singh S (2017) CFD analysis of solar air heater duct having square wave profiled transverse ribs as roughness elements. Sol Energy 162:442–453. https://doi.org/10.1016/j.solener.2018.01.019
15. Saedodin S, Zamzamian SAH, Nimvari ME, Wongwises S, Jouybari HJ (2017) Performance evaluation of a flat-plate solar collector filled with porous metal foam: experimental and numerical analysis. Energy Convers Manag 153(August):278–287. https://doi.org/10.1016/j.enconman.2017.09.072
16. Thakur DS, Khan MK, Pathak M (2017) Solar air heater with hyperbolic ribs: 3D simulation with experimental validation. Renew Energy 113:357–368. https://doi.org/10.1016/j.renene.2017.05.096

# Evaluation of Pollutants Emitted from Open Field Crop Residue Burning in Punjab, India

**Harpreet Singh Dhaliwal, Yadwinder Singh Brar, and Gursewak Singh Brar**

**Abstract** Air pollution is one of the biggest problems of the modern World. There are so many harmful impacts of air pollution, which affects the human life and vegetation in the direct or indirect manner. There are so many reasons of air pollution; open field crop residue burning is one of the main reasons of air pollution. Indian state Punjab is one of the most grain producing states in the country. Due to some social, economic and political reasons, the residues of crops are burnt by the farmers in this state. Burning of crop residues emit heavy air pollutants like $CO_2$, CO, $NO_x$, $SO_2$ and particulate matter. All these pollutants affect the human life in the form of diseases. If the crop residues must be utilized to generate electrical energy, it will provide a solution to atmospheric pollution as well as provide electrical energy in a sustainable manner. The proposed study predicts the crop residue generation by preparing NDVI model of Punjab, using Landsat8 bands. The NDVI model is combined with data of statistical abstract of Punjab. Surplus crop residues are predicted to evaluate the pollutant emissions from open field crop residue burning. District-wise pollutant emissions are evaluated in the study. It is proposed to generate electrical energy from these crop residues to limit the air pollution.

**Keywords** Crop residue burning · Pollutants · Biomass

## 1 Introduction

Clean air and clean water are the basic requirements for living beings on Earth. Due to increase in population, upliftment in living standards of people, rapid use of automobiles, rapid growth of industrialization and increased demand of energy, the atmospheric and water pollution has been increased in India. The problem of

H. S. Dhaliwal (✉) · Y. S. Brar
Inder Kumar Gujral Punjab Technical University, Kapurthala, Punjab, India

G. S. Brar
Baba Banda Singh Bahadur Engineering College, Fatehgarh Sahib, Punjab, India



R. Kumar et al. (eds.), *Recent Trends in Thermal Engineering*, Lecture Notes in Mechanical Engineering, https://doi.org/10.1007/978-981-16-3132-0_20

environmental pollution cannot be solved overnight; it will take years of positive practice to solve it. The air quality of India is considered to be one of the most polluted in the world [1]. There are so many reasons of elevated environmental pollution; one of them is burning of crop residues in the open fields. After the harvesting of every crop, heavy amount of crop residues is left in the fields. Farmers need to clear their fields in order to prepare the fields for the sowing of the next crop. Owing to no fruitful use and no sustainable policy for the management of these crop residues, the farmers of Northern India burn these crop residues in their fields. This malpractice is highly adopted by the farmers of Punjab, India [2]. The open field burning of crop residues gives rise to environmental pollution.

Many harmful pollutants are emitted from the burning of crop residues in the open fields. The major pollutant emitted from the burning of crop residue is $CO_2$. Besides this, CO, $SO_2$, $No_x$, particulate matter ($PM_{2.5}$ and $PM_{10}$), elemental and organic carbon is emitted from the burning of crop residues, which are the major reasons of developing respiratory diseases like asthma in the people [3, 4]. The authors have concluded from their study that children are at the highest risk to develop respiratory infections due to the crop residue burning [5, 6]. The harvesting of the paddy crop took place in the months of October and November in the Northern parts of India. The authors in their studies have reported that the atmospheric pollution has rapidly increased during the months of October and November every year [7, 8]. According to the author, one of the reasons behind this is burning of crop residues. According to studies, areas near the fields where crop residues are burnt are highly polluted [9, 10]. To reduce the rise in pollution, we have proposed the use of crop residue-based biomass power plants in Punjab state. It will help to decelerate the rise of pollutants in the air. It is also presented that the controlled burning of crop residues emits lesser pollutants as compared to open burning of them on fields.

## 2 Materials and Methods

In this study, lansat8 satellite data of Punjab is taken from the US Geological Survey (USGS) [11]. The satellite data is used to predict the Normalized Difference Vegetation Index (NDVI) model of state and district-wise crop residue generation is calculated from this data. Surplus crop residues are predicted and evaluated for the generation of pollutant emissions from the burning of crop residues in each district of the state.

### *2.1 Area of Study*

Punjab is the Northern state of India. This state is one of the highest crops producing states of India. Area of 50,362 $Km^2$ falls under the jurisdiction of this state

[12]. Punjab holds nearly 1.53% of India's whole area. Punjab is full of variety when it comes to terrain and weather. The Punjab region has both planes and mountains within its territory. Punjab has varied weather conditions. In summers, temperature ranges up to 48 °C, whereas in the extreme winters, this region observes an extreme fall in temperature to 0 °C. Punjab receives a fair amount of rainfall per year which also helps in making soil more fertile. Major crops of Punjab are wheat, rice and cotton [12]. Punjab is the largest contributor of food grains in the central pool.

## 2.2 GIS Data Source

The data used to predict emitted aerosols (pollutants) is taken from USGS, which is a good source of satellite images in eleven bands [11]. One of the major parameters for the estimation of the pollutants is to predict the crop and the non-crop area. Crop area and non-crop area can be predicted by developing Normalized Difference Vegetation Index (NDVI) model of the state, which is prepared by using Landsat8 5th band (near infrared band) and the 4th band (visible red band) as given in Eq. (1). This study uses data having cloud interference lesser than 10% from USGS. It is followed by orientation of the bands by 12.8° to make it vertically aligned. Threshold value of NDVI to predict crop and non-crop area is set at 0.3529.

$$\mathrm{NDVI} = \frac{\mathrm{NIR} - \mathrm{Red}}{\mathrm{NIR} + \mathrm{Red}} \tag{1}$$

$$\begin{aligned} \mathrm{NDVI} = \text{if } \ \mathrm{ndvi\_raw}_{(i,j)} > 0.3529 \ \text{ then } \ & \mathrm{NDVI}_{(i,j)} = 1 \\ \text{else } \ & \mathrm{NDVI}_{(i,j)} = 0 \end{aligned} \tag{2}$$

where $i$ is longitude and $j$ is the latitude of locations (Fig. 1).

## 2.3 Crop Residue Production, Local Usage and Surplus Crop Residue

Crop residues are the materials left behind in the fields, after the harvesting of crops. These crop residues are generally in the form of straw, husk and stalks which are found in the crops grown in Punjab. To calculate the potential of surplus crop residual, first we must have to calculate the gross crop residues produced after the harvesting of crops. Gross crop residues are calculated by multiplying CRR (crop to residue ratio) values with grain production. The grain production in the state has been taken from the statistical abstract of Punjab [12]. The values of CRR taken

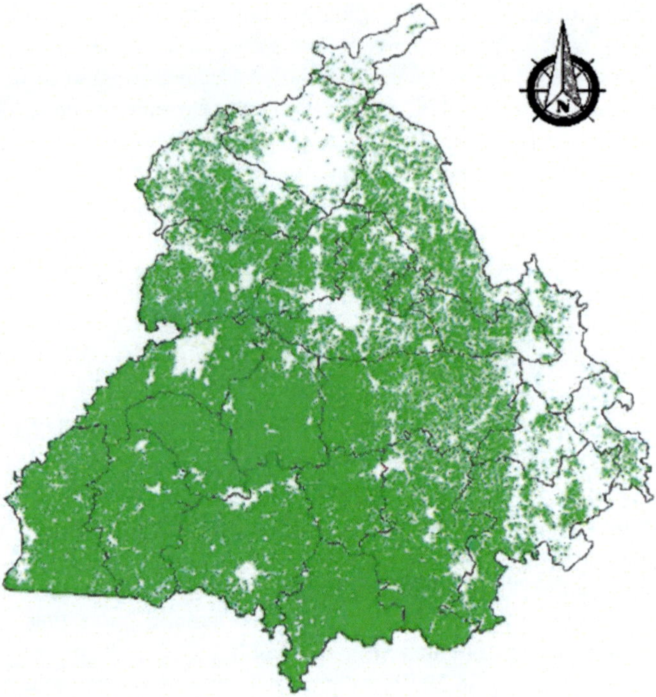

**Fig. 1** Crop and non-crop area of Punjab

from the literature [13]. The gross crop residue ($GCR_i$) production from different crops, as given by Eq. (3)

$$GCR_i = GP_i \times CRR_i \quad (3)$$

*where* $GCR_i$ *is the gross crop residue of ith crop,* $GP_i$ *is the grain production of ith crop,* $CRR_i$ *is the average crop residue of ith crop.*

In this study, major crops like rice, wheat and cotton are taken into account to predict the amount of crop residual. There are many ways in which crop residue is used for other purposes such as fodder, fuel wood, thatching, manuring and household fuel. The crop residue used for local purposes ($CRNP_i$) is calculated by using Eq. (4).

$$CRNP_i = F_i + FW_i + T_i + M_i + SF_i + MS_i \quad (4)$$

*where* $F_i$ *represents crop residues used for fodder making,* $FW_i$ *represents crop residue used as fuel wood,* $T_i$ *represents the use of crop residue for thatching,* $M_i$ *represents the use of crop residues for manuring,* $SF_i$ *represents the crop residue sown in fields and* $MS_i$ *represents the crop residue used for miscellaneous purposes.*

The values of crop residues used to meet local requirements are taken from the literature [13]. It is found that approximately, 66.36% of crop residues are used in above mentioned local purposes. The crop residues left after the local uses are considered as surplus crop residues, which are burnt by the farmers in their fields. Surplus crop residues ($SCR_i$) are calculated by using Eq. (5)

$$SCR_i = GCR_i - CRNP_i \tag{5}$$

It has been found from the calculation work that as per statistical abstract of Punjab, the gross grain production in the state for the year 2017–18 was 41.031 Million tonnes. The gross crop residue produced from this grain production was 43.812 Million tonnes. Out of the gross crop residue produced, 29.074 Million tonnes were utilized in local purposes. The surplus crop residue, which is burnt in the open fields of state comes out to be 14.738 Million tonnes, which is around 33.64% of the total crop residue generated in the year 2017–18.

## 2.4 *Burning of Crop Residues in the Fields and Emission of Pollutants*

There are many types of pollutants which are emitted through various sources. This study focuses only on the pollutants emitted from crop residue burning. Table 1 shows emission factors of three major crop residues, which are emitted from the open field burning of crop residues. These pollutant emissions are responsible for air pollution as well as for the depletion of ground water of the state. Change in climate is a result of increased air pollution and studies have also shown that $CO_2$ along with various other gases like CO, sulfur dioxide and nitrogen oxide contributes to the greenhouse effect, which is further responsible for global warming, acid rain and climate changes of the planet.

**Table 1** Emission factors of crop residues burning (g/kg) [14]

| Emission factor | TSP | $PM_{10}$ | EC | OC | $So_2$ | $No_x$ | CO | $CO_2$ |
|---|---|---|---|---|---|---|---|---|
| Wheat straw | 12.00 | 05.74 | 01.20 | 02.20 | 0.47 | 2.33 | 66.69 | 1194.88 |
| Rice straw | 05.30 | 03.46 | 0.860 | 0.94 | 0.62 | 2.84 | 31.39 | 1162.15 |
| Cotton stalk combined | 11.00 | – | 0.690 | 05.00 | 0.40 | 2.50 | 58.00 | 1515.00 |

*TSP* Total suspended particles, $PM_{10}$ pariculate matter 10 μ, *EC* elemental carbon, *OC* organic carbon, $SO_2$ sulfur dioxide, $No_x$ nitrogen oxides, *CO* carbon monoxide, $CO_2$ carbon dioxide

To compare the various pollutants with and without burning of crop residues in the open field, it is important to calculate the quantity of emitted pollutants. Pollutant emissions were calculated using Eq. (5) for every district of study area [15].

$$\text{poll\_em}_{(i,j)} = \sum_{k=1}^{k=m} \text{BM}_k \times \text{EF}_{(i,j)} \tag{6}$$

where BM represents biomass of burnt crop in *k*th district, m is the number of geographical locations in *k*th district

EF $_{(i,j)}$ is the emission factor of *i*th crop of *j*th pollutant (g/kg).

It is observed from Eq. (6) that emission of pollutants depends upon the emission factors and the amount of burning of the crop residues in open field. District-wise surplus crop residues have been taken into account from Eq. (5). Each district is evaluated for the pollutant emissions. Table 2 shows district-wise emission factors of each pollutant in each district of the state. Emission factors for pollutants like $CO_2$, CO and TSP are the maximum among other pollutants in this group. Therefore, it is very much important to reduce the emission rate of these gasses as much as possible.

## 3 Results and Discussions

The study used Landsat8 bands to predict the NDVI modelto distinguish the crop and the non-crop area of the state of Punjab, INDIA. NDVI calculations are followed by the prediction of crop residues in 22 districts of Punjab. The burning of biomass in open fields will emit harmful aerosols and the respective calculations are shown in Table 2. The results show that total 6608.5 kt of atmospheric pollution has been generated from the open field burning of crop residues in Punjab. Total emission of suspended particles is 47.73 kt from the crop residue burning in the air of the state, while 269.17 kt of CO, 13.38 kt of nitrogen oxides, 5.3 kt of elemental carbon and 9.11 kt of organic carbon have been added to the air of state. The biggest contributor of GHG is $CO_2$, which amounts for 6236.85 kt from the crop residue burning in the state. It has been observed from the results that district Barnala is the most polluted district in the state, whereas Pathankot is the least polluted district in the state. It is also observed that the districts Sangrur, Ferozepur, Ludhiana and Bathinda also contribute heavily in atmospheric pollution of the state. This study proposed that the setting-up of biomass power plants in the state will reduce the rate of increasing pollutants in the atmosphere (Fig. 2).

**Table 2** Harmful pollutants by crop residue burning in Punjab (District-wise) [12]

| | Pollutant emissions in kilo tonnes | | | | | | | | |
|---|---|---|---|---|---|---|---|---|---|
| Districts | TSP | $PM_{10}$ | EC | OC | $So_2$ | $No_x$ | CO | $CO_2$ | Total |
| Fazilka | 2.0970 | 0.8499 | 0.2065 | 0.4932 | 0.0993 | 0.5013 | 11.6250 | 252.602 | 268.4742 |
| Ferozepur | 2.8024 | 1.4795 | 0.3208 | 0.5090 | 0.1717 | 0.8163 | 15.8654 | 374.408 | 396.3731 |
| Shri Muktsar Sahib | 2.6741 | 1.2666 | 0.2867 | 0.5479 | 0.1464 | 0.7126 | 14.9963 | 340.009 | 360.6396 |
| Faridkot | 2.6348 | 1.3767 | 0.2994 | 0.4840 | 0.1592 | 0.7585 | 14.9006 | 349.381 | 369.9942 |
| Amritsar | 2.0926 | 1.0934 | 0.2362 | 0.3804 | 0.1231 | 0.5873 | 11.8232 | 271.805 | 288.1412 |
| Taran Tarn | 2.2412 | 1.1841 | 0.2568 | 0.4070 | 0.1377 | 0.6545 | 12.6901 | 300.036 | 317.6074 |
| Bathinda | 2.7502 | 1.1466 | 0.2789 | 0.6424 | 0.1418 | 0.7074 | 15.3036 | 348.423 | 369.3939 |
| Moga | 2.6548 | 1.4116 | 0.3068 | 0.4818 | 0.1672 | 0.7931 | 15.0507 | 361.612 | 382.478 |
| Gurdaspur | 2.0003 | 1.0450 | 0.2257 | 0.3637 | 0.1176 | 0.5611 | 11.3014 | 259.704 | 275.3188 |
| Kapurthala | 2.0921 | 1.1144 | 0.2423 | 0.3796 | 0.1326 | 0.6287 | 11.8650 | 286.288 | 302.7427 |
| Jalandhar | 2.0336 | 1.0709 | 0.2320 | 0.3694 | 0.1234 | 0.5872 | 11.5076 | 269.878 | 285.8021 |
| Mansa | 2.6941 | 1.1925 | 0.2797 | 0.5927 | 0.1415 | 0.6982 | 15.0398 | 339.570 | 360.2085 |
| Barnala | 2.8027 | 1.4709 | 0.3214 | 0.5171 | 0.1742 | 0.8289 | 15.8705 | 379.626 | 401.6117 |
| Pathankot | 0.9903 | 0.5071 | 0.1088 | 0.1804 | 0.0537 | 0.2578 | 05.5739 | 121.574 | 129.246 |
| Ludhiana | 2.2200 | 1.1794 | 0.2563 | 0.4029 | 0.1393 | 0.6611 | 12.5840 | 301.653 | 319.096 |
| Hoshiarpur | 1.0082 | 0.5136 | 0.1099 | 0.1837 | 0.0534 | 0.2571 | 5.6687 | 121.912 | 129.7066 |
| Sangrur | 2.7851 | 1.4625 | 0.3189 | 0.5120 | 0.1721 | 0.8185 | 15.7676 | 375.271 | 397.1077 |
| SBS Nagar | 1.8273 | 0.9545 | 0.2062 | 0.3322 | 0.1074 | 0.5123 | 10.3241 | 237.164 | 251.428 |
| Patiala | 2.3577 | 1.2381 | 0.2684 | 0.4296 | 0.1424 | 0.6781 | 13.3371 | 312.089 | 330.5404 |
| Fatehgarh Sahib | 2.4234 | 1.2802 | 0.2776 | 0.4401 | 0.1488 | 0.7075 | 13.7216 | 324.338 | 343.3372 |
| Rupnagar | 1.3279 | 0.6801 | 0.1459 | 0.2419 | 0.0720 | 0.3457 | 7.4740 | 163.028 | 173.3155 |
| SAS Nagar | 1.2248 | 0.6216 | 0.1329 | 0.2233 | 0.0639 | 0.3077 | 6.8820 | 146.474 | 155.9302 |
| Total | 47.735 | 24.140 | 05.319 | 09.115 | 02.789 | 13.382 | 269.173 | 6236.855 | 6608.508 |

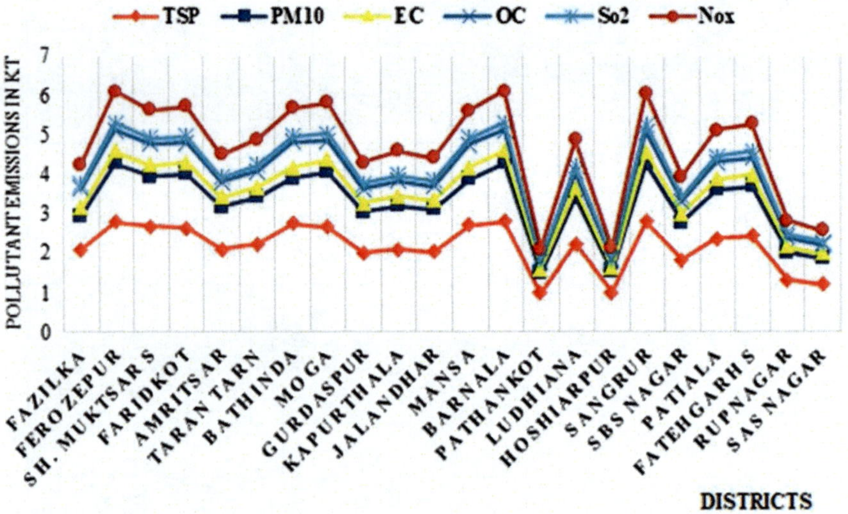


**Fig. 2** District-wise pollutant emissions in Punjab

## 4 Conclusions

This study was aimed to compute the atmospheric pollution produced in Punjab state from the burning of open field crop residues. It is concluded from the study that heavy amount of pollutants has been emitted from the burning of crop residues in the state. These pollutants are mainly responsible for the development of serious respiratory diseases in the people of state. Heavy amount of $CO_2$ has been released, which is mainly responsible for the increase in GHG and global warming. It is also concluded that the districts, which produces high grains, are more effected from the pollution of crop residue burning. The best possible way to reduce the pollution of crop residue burning is to set up biomass power plants in the state, which will generate electrical power for the sustainability of the state as well as help to control the atmospheric pollution.

## References

1. Shyamsundar P et al (2019) Fields on fire: alternatives to crop residue burning in India. Science 365(6453):536–538
2. Gupta N (2019) Paddy residue burning in Punjab understanding farmers' perspectives and rural air pollution. CEEW 1–44
3. Satyendra T, Singh RN, Shaishav S (2013) Emissions from crop/biomass residue burning risk to atmospheric quality. Int Res J Earth Sci 1(1):1–5

4. Ravindra K, Singh T, Mor S (2019) Emissions of air pollutants from primary crop residue burning in India and their mitigation strategies for cleaner emissions. J Cleaner Prod 236:261–273
5. Gupta S, Agarwal R, Mittal SK (2016) Respiratory health concerns in children at some strategic locations from high PM levels during crop residue burning episodes. Atmos Environ 137:127–134
6. Chakrabarti S, Khan MT, Kishore A, Roy D, Scott SP (2019) Risk of acute respiratory infection from crop burning in India: estimating disease burden and economic welfare from satellite and national health survey data for 250,000 persons. Int J Epidemiol 48(4):1–12
7. Singh J, Panesar BS, Sharma SK (2010) A mathematical model for transporting the biomass to biomass based power plant. Biomass Bioenergy 34(4):483–488
8. Jethva H, Chand D, Torres O, Gupta P, Lyapustin A, Patadia F (2018) Agricultural burning and air quality over northern India: a synergistic analysis using nasa's a-train satellite data and ground measurements. Aerosol Air Qual Res 18(7):1756–1773
9. Koutsias N, Balatsos P, Kalabokidis K (2014) Fire occurrence zones: Kernel density estimation of historical wildfire ignitions at the national level, Greece. J Maps 10(4):630–639
10. Koutsias N, Kalabokidis KD, Allgöwer B (2004) Fire occurrence patterns at landscape level: beyond positional accuracy of ignition points with kernel density estimation methods. Nat Res Model 17(4):359–375
11. USGS homepage, https://earthexplorer.usgs.gov/. Last accessed 2020/02/17
12. Statistical abstract of Punjab-2018. Published by Economic Adviser to Government of Punjab. Publication No. 958, 1-765
13. Online available https://www.esopb.gov.in/static/PDF/Abstract2018.pdf
14. Chauhan S (2012) District wise agriculture biomass resource assessment for power generation: a case study from an Indian state, Punjab. Biomass Bioenergy 37:205–212
15. Cao G, Zhang X, Gong S, Zheng F (2008) Investigation on emission factors of particulate matter and gaseous pollutants from crop residue burning. J Environ Sci 20(1):50–55

# Analysis on Energy-Efficient HVAC System for Buildings

**Neelesh Patel and D. Buddhi**

**Abstract** HVAC system plays a very important role for the building, as it has captured almost 45–50% of the electrical load of any building. This study shows that how we can make sure conventional HVAC to energy-efficient HVAC system. This study also shows that it will also reduce the carbon footprint as electrical load going down.

**Keywords** HVAC · AHU · Carbon emission reduction

## 1 Introduction

Building energy consumption is the major concern nowadays. In 2009, global public building energy consumption was 2 billion TCE, representing 11.4% of total building consumption [1]. The important factor is that office building is almost one-fifth of total building energy use [2].

Air conditioning is 30–40% of the total building energy consumption of office building [3]. Climate is also playing important role in case of air-conditioning energy consumption. The potential impacts on the various types of weather forecast models, weather data and building prototypes have been studied from varied prospective [4–8].

A HVAC system schematic diagram is shown in Fig. 1.

In HVAC, evaporator, condenser, compressor and expansion valve are the main components; principally, evaporator and condenser are the heat exchangers, one circuit for refrigerant acts as a primary circuit, and refrigerant absorbs the heat from return heating load (chilled water return line) and releases on cooling tower side in condenser.

N. Patel (✉)
Center of Excellence in Renewable & Sustainable Energy Studies,
Suresh Gyanvihar University, Jaipur, India

D. Buddhi
School of Mechanical Engineering, Lovely Professional University, Phagwara, India



R. Kumar et al. (eds.), *Recent Trends in Thermal Engineering*, Lecture Notes in Mechanical Engineering, https://doi.org/10.1007/978-981-16-3132-0_21

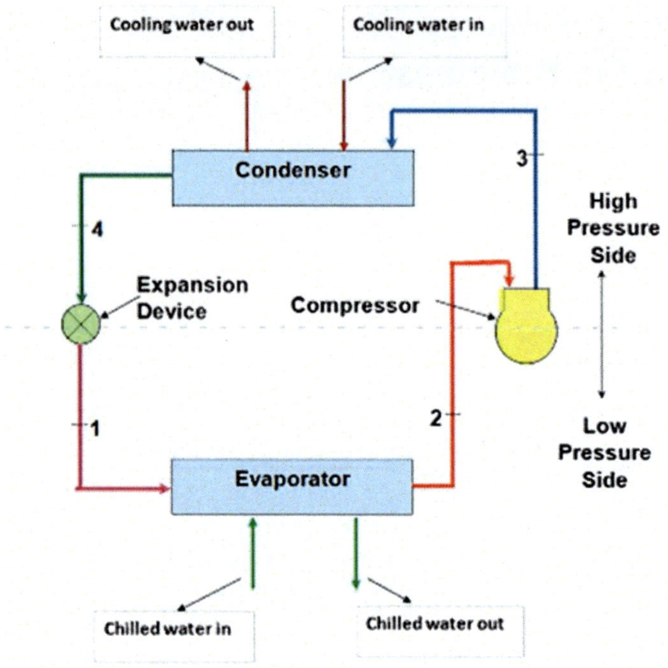


**Fig. 1** HVAC schematic diagram

All equipment come under high side of equipment where it will generate the chilled water.

Chilled water and AHU circuit—This circuit is normally known as low side of HVAC system, where generated chilled water has been utilized (Fig. 2).

Standard energy distribution % for building energy consumption clearly shows that maximum consumption has been for chiller machine.

**Fig. 2** Chilled water and AHU circuit

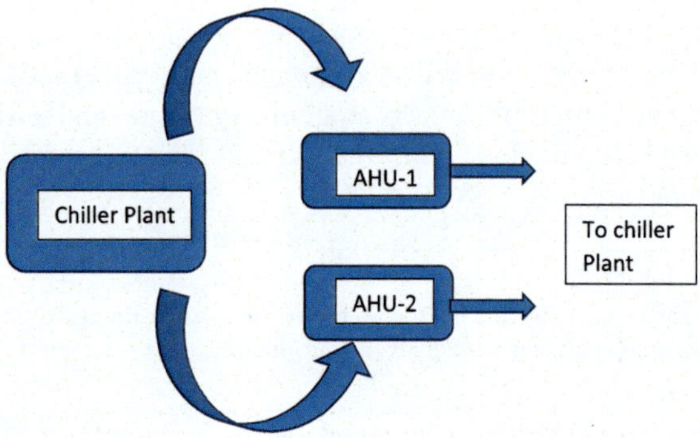

**Table 1** HVAC energy distribution

| Name | % Consumption (%) |
|---|---|
| Chiller | 48 |
| Primary pump | 1 |
| Secondary pump | 2 |
| Condenser pump | 2 |
| CT | 1 |
| AHU | 14 |
| Others | 32 |

Table 1 and Fig. 3 have clearly shown that % distribution of HVAC system maximum % consumption has been taking care by chiller machine which is 48%, pumping system has been 5%, and air handling system consumption is close to 14%.

% Reduction on energy consumption through energy conservation measures—

1. Using VFD for chilled water pump;
2. Using VFD for condenser pump;
3. Using On/Off control for cooling tower fan;
4. Using VFD for AHUs with RT air feedback.

Doing all energy conservation measures, the direct impact on chiller machine has been shown.

Working principle of improved infrastructure of HVAC system with energy conservation measures—There are two distributions: One is high side, and other one is low-side improvement on low side with the use of VFD on AHU fan as per heat load inside the building; accordingly, chilled water flow has been controlled

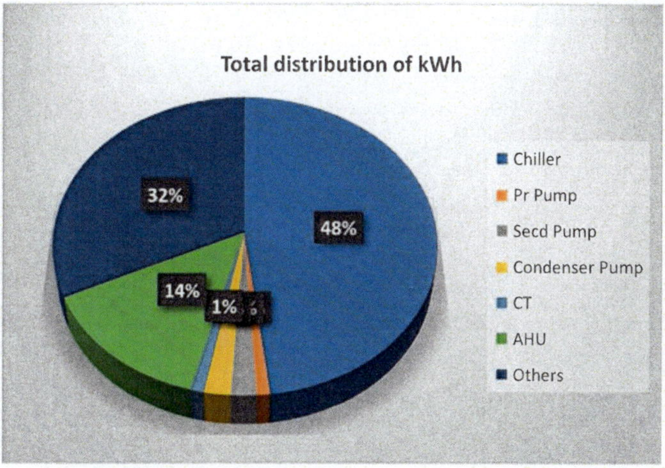


**Fig. 3** HVAC energy distribution

through actuator valve on secondary pump; on that reference secondary pump, VFD reduces the speed of pump and at the same time kWh consumption also goes down on both AHU and pumping system.

Now the measured impact comes on chiller machine because when the heating load on building has been less, all variations taking place, and at the same time, generation of chilled water is also going down.

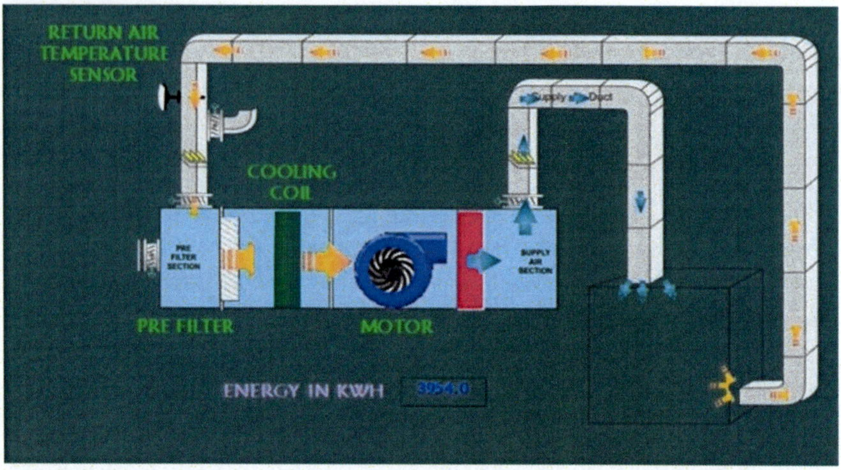


% Reduction on energy consumption in HVAC system (Fig. 4; Table 2).

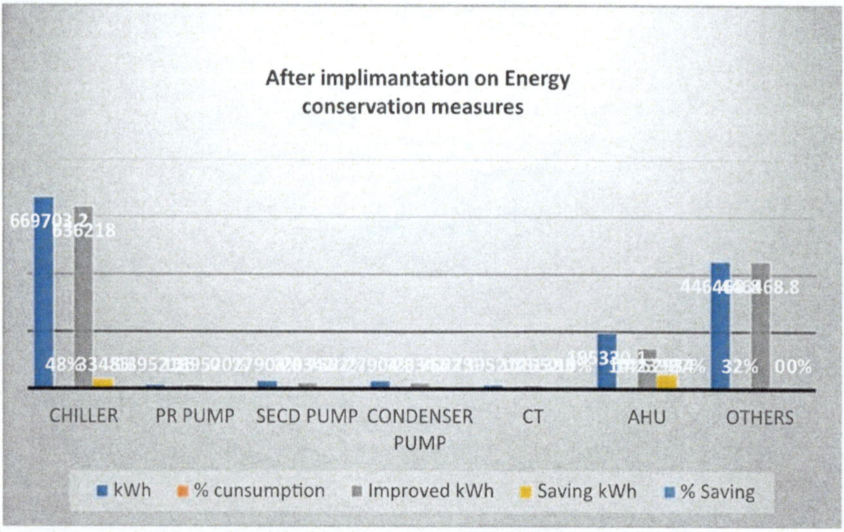


**Fig. 4** Reduced energy consumption after implementation of energy conservation measures

**Table 2** % Saving in kWh

| Name | % Consumption (%) | % Saving (%) |
|---|---|---|
| Chiller | 48 | 5 |
| Primary pump | 1 | 0 |
| Secondary pump | 2 | 27 |
| Condenser pump | 2 | 27 |
| Cooling tower (CT) | 1 | 10 |
| AHU | 14 | 27 |
| Others | 32 | 0 |

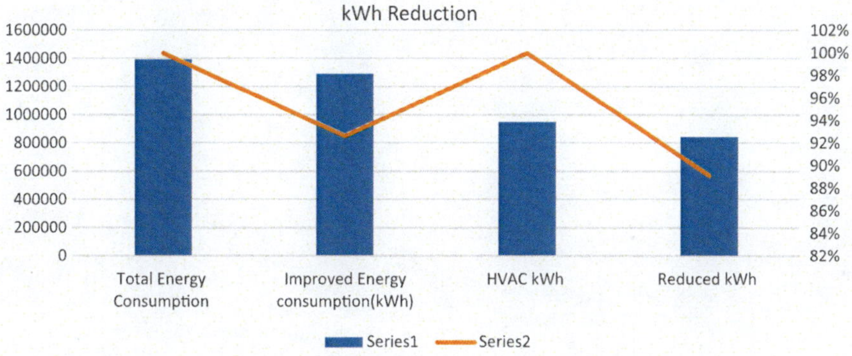


**Fig. 5** % Reduction on energy consumption on totality and individual HVAC system

# 2 Conclusion

Total energy consumption—1,395,215 kWh (100%);
Reduced energy consumption—1,292,276.2 kWh (93%);
Saving in energy on total bill—102,938.96 kWh (7%);
On HVAC energy consumption reduction—845,807 kWh (11%).

After implementation on energy conservation measures, % reduction in totality is 7%, and in separate HVAC, reduction is 7% (Fig. 5).

# References

1. US Energy Information Administration (2010) International energy outlook 2010[R/OL]. http://www.eia.gov/oiaf/ieo/index.html
2. Pérez-Lombarda L, Ortizb J, Poutb C A review on buildings energy consumption information

3. Building energy research center in Tsinghua University, 2013 Annual Report on China Building Energy Efficiency, China Building Industrial Publishing, Beijing, 2013 (in Chinese)
4. Long N (2006) Real—time weather data access guide. User's guide, NREL/BR-550-34303
5. Crawley DB Estimating the impacts of climate change and urbanization on building performance US Department of Energy, Washington, DC, USA
6. Wilcox S, Marion W (2008) Users manual for TMY3 data sets, Technical report NREL/TP-581-43156, NREL
7. Hong T, Chang W-K, Lin H-W A fresh look at weather impact on peak electricity demand and energy use of buildings using 30-year actual weather data
8. Cui Y, Yana D, Hong T, Xiao C, Luo X, Zhang Q Comparison of typical year and multiyear building simulations using a 55-year actual weather data set from China

# Optical and Thermal Analysis of STCR Cavity Subjected Under Flow and no Flow Conditions

**Jeet Prakash Sharma, Ravindra Jilte, and Ravinder Kumar**

**Abstract** Solar thermochemical water/$CO_2$-splitting process is used to produce hydrogen and syngas. Since past 50 years, solar thermochemical processes have been explored and advanced on the laboratory scale. At current stage, cerium oxide ($CeO_2$) catalyst-based water-splitting process seems more promising than any other. Further, there has been significant advancement in the area of solar receivers and solar simulators. Thus, this study aims to explore the solar flux distribution in the RPC-based solar thermochemical reactor-receiver cavity. This paper presents the conceptual design and optical analysis of RPC cavity to attain the optimum conditions (uniform distribution of solar flux) and to achieve the maximum temperature for reduction reaction. In this study, a hybrid geometry of solar reactor-receiver cavity (cylindrical and hemi-spherical) has been used. The focal point of the simulator has moved [(1) on the cavity aperture and (2) 20 mm inside the cavity] to analyze the solar flux and temperature distribution inside the cavity. The distribution of solar flux inside the solar thermochemical reactor cavity found to be more uniform for case-1 with slope error 4. It was seen that case-1 with slope error 2 mrad yields 33% higher solar flux value as compared to the 4 mard slope error and yet 4 mrad slope error yield the higher temperature inside reactor cavity.

**Keywords** Solar reactor-receiver cavity · Solar flux distribution · Optical analysis · Hydrogen/syngas production · Water-splitting process

J. P. Sharma
School of Mechanical Engineering, Lovely Professional University, Phagwara, Punjab 144411, India

R. Jilte (✉) · R. Kumar
Department of Mechanical Engineering, Lovely Professional University, Phagwara, Punjab 144411, India

 
R. Kumar et al. (eds.), *Recent Trends in Thermal Engineering*, Lecture Notes in Mechanical Engineering, https://doi.org/10.1007/978-981-16-3132-0_22

## 1 Introduction

The rapid depletion of fossil fuels and catastrophic shift in the climate are closely related. Overconsumption of conventional energy resources has damaged the environment. Thus, researchers have shifted their focus on the advancement, development and pilot scale production of solar fuels. Renewable energy is the control factor for the safety of environment, and at this point, the maximum utilization of renewable energy is critical. Among all the renewable energy sources, solar energy happens to be best suited due to its abundance, inexhaustibility and usage flexibility from industry to home. Solar thermochemical process are based on the concept of conversion of solar energy into fuels (hydrogen and syngas) via. water or carbon di oxide splitting process. Among these five solar thermochemical processes, namely solar thermolysis, solar thermochemical cycles, solar gasification, solar cracking and solar reforming, solar thermochemical cycles are widely adopted and experimented around the world for hydrogen and syngas production. Solar thermolysis is a process where water is dissociated into hydrogen and oxygen molecules $\left[H_2O \rightarrow H_2 + \frac{1}{2}O_2\right]$ by supplying the enormous amount of heat until $\Delta G_{\text{reaction}} \leq 0$. Since this process requires very high temperature (up to 4300 K) to carry out the water dissociation reaction thus, it is not a viable option [1]. Methane reforming is a carbon feed-based solar thermochemical process which yields syngas [2]. These reactions given below take place at the temperature of 923–1473 K.

$$CH_4 + H_2O \rightarrow 3H_2 + CO \quad \Delta H = 206\,\text{KJ/mol}$$
$$CH_4 + CO_2 \rightarrow 2H_2 + 2CO \quad \Delta H = 247\,\text{KJ/mol}$$

Solar cracking is another solar thermochemical process where methane is dissociated into carbon and hydrogen $[CH_4 \rightarrow C + 2H_2]$ at the temperature of 800–1500 K [3]. Solar gasification is used to produce syngas by steam gasifying the carbon feed stock at the temperature of 1123–1883 K [4]. The gasification reaction is given below;

$$C_aH_bO_c + (a-c)H_2O \rightarrow \left(\frac{b}{c} + a - c\right)H_2 + aCO$$

Thermochemical is the best suited option to produce hydrogen and syngas because it resolved the high temperature and hydrogen–oxygen separation problem encountered in solar thermolysis process. Solar thermochemical cycles are classified into two-step and multi-step cycles. Currently, ceria ($CeO_2$)-based solar thermochemical has been opted as the best available candidate for hydrogen production via two-step water or $CO_2$ splitting cycle. The thermal reduction and re-oxidation of ceria (by $H_2O$ and $CO_2$) has been given in the below equation;

Reduction reaction:

$$CeO_2 \rightarrow CeO_{2-\delta} + \frac{\delta}{2} O_2$$

Re-oxidation reaction:

$$\text{By water:} \quad CeO_{2-\delta} + \delta H_2O \rightarrow CeO_2 + \delta H_2$$
$$\text{By carbon di oxide:} \quad CeO_{2-\delta} + \delta CO_2 \rightarrow CeO_2 + \delta CO$$

The nonstoichiometric ceria-based compound has been widely investigated [5, 6]. In 2006, first study of ceria-based reactor used for the production of hydrogen was published. After that, ceria has been a prime candidate for solar thermochemical cycle and researchers around the world have experimented and investigated it thoroughly. Studies have examined the reduction capacity and fuel production capability of ceria over the wide ranges of temperature. Researchers have attempted to reduce the gap between reduction and re-oxidation step of ceria-based redox cycle. Many studies have been carried out to investigate the thermal performance of cavity receivers of different shape and sizes. Jilte et al. carried out studies on different shapes (conical, reverse conical, cylindrical, spherical, and hetro-conical) of cavity receivers and compared their thermal performance [7–9]. Results of their study revealed that the conical cavity has lowest thermal (convective) losses among all other cavity shapes. The tabular cylindrical shaped cavity receiver was also analyzed for the understanding the effects of operational and structural parameters on thermal performance [10]. A review on volumetric receiver revealed that the 12% of input energy is lost as convective and radiative losses at the aperture window of solar reactor-receiver cavity [11]. The optical analysis for CSP system is performed using the Monto Carlo Ray Tracing (MCRT) method [12]. Study carried out by Li et al. proved that the for a parabolic dish system, optical efficiency is a function of focal length and aperture diameter [13]. Daabo et al. performed an investigation on three receiver geometries using MCRT and results showed that the value of solar flux depends on the type of cavity geometry [14, 15]. Daabo et al. carried out another study to investigate and compare the optical efficiencies of three cavity shapes (cylindrical, spherical and conical) [14, 16–18].

## 2 Conceptual Design of STCR Cavity

In this present study, a solar thermochemical reactor cavity is designed for the geometrical and optical analysis as shown in Fig. 1. The solar thermochemical reactor cavity has a quartz glass window for incoming solar flux. The RPC catalyst thickness is kept 20 with 5 mm gas flow gap. To carry out this analysis, the study has divided into two cases. In case-1, the focal point is kept at the cavity aperture and in case-2 the focal point is kept 20 mm inside the reactor cavity as shown in Fig. 2. For both cases, the slope error is varied from 2 mrad to 4 mrad to obtain the flux profile using SolTrace. This flux profile is used to obtain the flux and temperature distribution on the reactor cavity walls using the CFD software Fluent v16.0. The DO radiation model is used to solve the radiative transport equation for the numerical analysis.

## 3 Mathematical Modeling

The governing equations used in the present simulations are given as follows:

(i) **Continuity equation**:

$$\frac{\partial \rho}{\partial t} + \nabla.(\rho \vec{v}) = 0 \quad (1)$$

where $\rho$ is fluid density and $\vec{v}$ is superficial fluid velocity.

(ii) **Momentum equation**:

$$\frac{\partial}{\partial t}(\rho \vec{v}) + \nabla.(\rho \vec{v}\vec{v}) = -\nabla p + \nabla.(\mu \nabla \vec{v}) \quad (2)$$

(iii) **Energy equation**:

$$\frac{\partial}{\partial t}(\rho h - p) + \nabla.[\vec{v}(\rho h)] = \nabla.(k\nabla T) + S_h \quad (3)$$

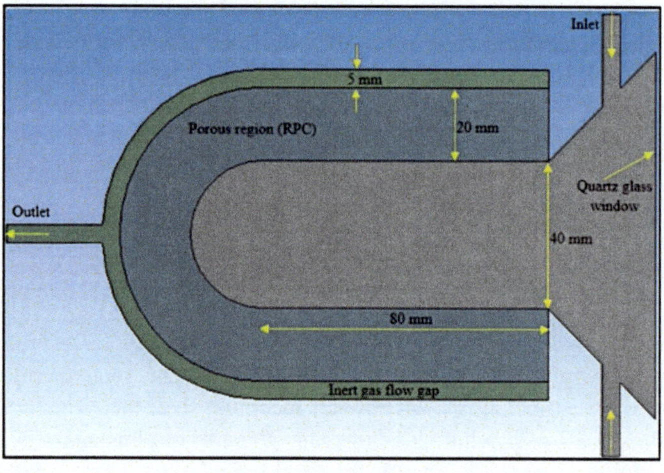


**Fig. 1** Schematic of solar thermochemical reactor cavity

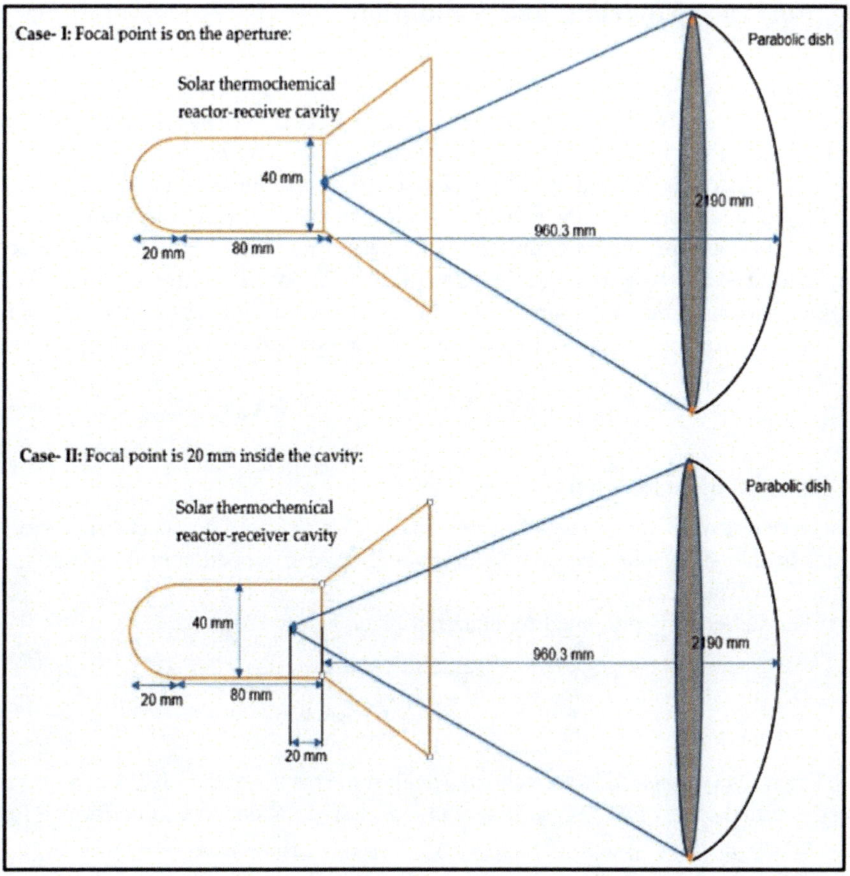


**Fig. 2** Schematic of two cases considered for optical analysis

For solid zones, Eq. (3) becomes,

$$\frac{\partial}{\partial t}(\rho h) = \nabla.(k\nabla T) \tag{4}$$

(iv) **Radiation transport equation**:

$$\begin{aligned} &\frac{\mathrm{d}I_\lambda(\vec{r},\vec{s})}{\mathrm{d}s} + (k_\lambda + \sigma_\lambda)I_\lambda(\vec{r},\vec{s}) \\ &= \varepsilon_\lambda I_{b,\lambda} + \frac{\sigma_\lambda}{4\pi}\int_0^{4\pi} I_\lambda(\vec{r},\vec{s'})\phi_\lambda(\vec{s}\rightarrow\vec{s'})\mathrm{d}\Omega' \end{aligned} \tag{5}$$

The radiative transfer equation is solved in the wavelength of 0 to 1.5 μm and 1.5 to ∞ μm.

## 4 SolTrace Modeling and Validation

SolTrace is Monto Carlo Ray Tracing (MCRT)-based software developed by National Renewable Energy Laboratory (NREL). This software is used for the optical modeling and analysis of solar systems [19]. The concentration performance of a CSP system is reduced by means of assorted errors. Since the surface area of a CSP system is quite large, the related errors follow the Gaussian distribution or bell curve. SolTrace software is embedded with the facility which allow users to select the study specific type of error, i.e., Gaussian or Pillbox. The error in the surface normal vector is known as slope error. The imprecise positioning of optical system leads to the tracking error. The total error can be denoted by Eq. (6) [20];

$$\sigma^2 = \sigma^2_{\text{specularity}} + \sigma^2_{\text{slope}} + \sigma^2_{\text{shape}} + \sigma^2_{\text{tracking}} \cong \sigma^2_{\text{slope,effective}} \tag{6}$$

$\sigma$ Standard deviation (SD)

SolTrace software combines the mirror and tracking errors into specularity errors [21]. In this study, the specularity error = 0.2 mrad has been considered for the optical analysis [22].

The total error is presented by equation given below [23].

$$\sigma_{\text{optical}} = \sqrt{\left(4\sigma^2_{\text{slope}} + \sigma^2_{\text{specularity}}\right)} \tag{7}$$

The results of this study are validated against the results of the MCRT code from study carried out by Lee [24] as shown in Fig. 3. The validation was performed for parameters given as follows;

Focal length of parabolic concentrator = 5 mm
Rim angle of parabolic concentrator = 45°
Slope error = 2 mrad

In the given Fig. 3a, SolTrace flux distribution has been validated against the publish study by Lee [24], and the results of the present study are very good agreement with the Lee's data. In Fig. 3b, c, SolTrace modeling of solar concentrator and STCR is presented and flux distribution is obtained.

## 5 Results and Discussion

Figure 4 shows that the flux obtained for case-1, where the focal point of the dish is kept at the cavity aperture ($f$ = 0.9803 m), yields the highest value for 2 mrad slope error. However, for same slope error, the flux for case-2, where the focal point is kept 20 mm inside the cavity, yield relatively and considerably low value.

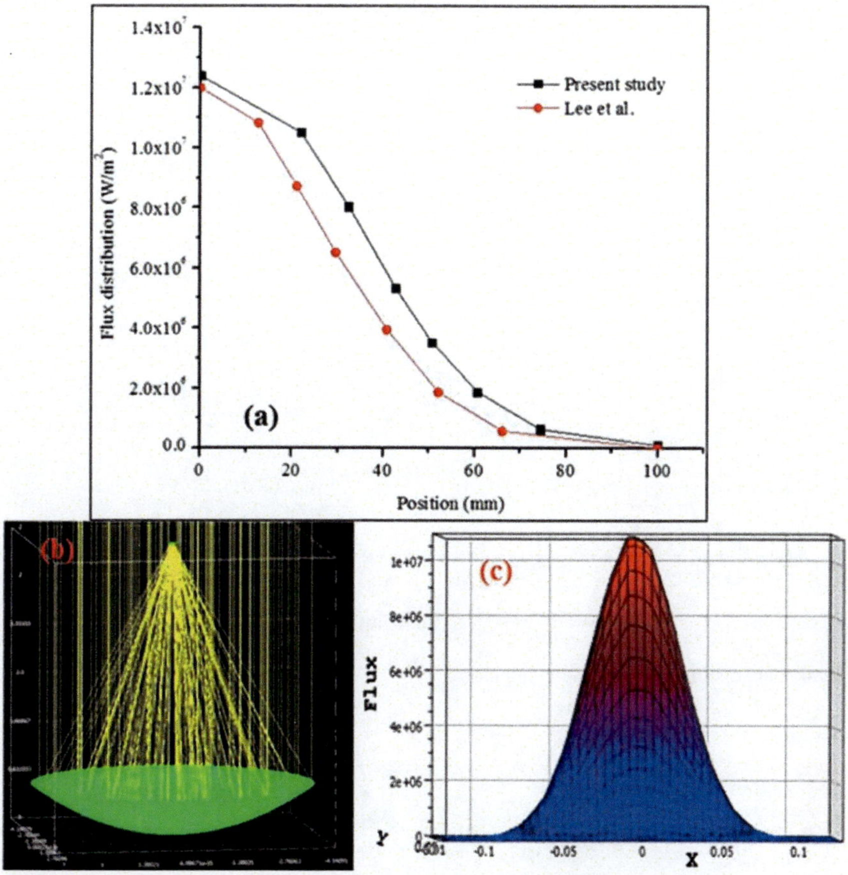


**Fig. 3** Flux distribution comparison and SolTrace validation with Lee's study [24]

To attain the uniform temperature distribution inside the reactor, cavity is main point of concern. Figure 5 shows the temperature contours for case-1 and case-2 under $N_2$ flow and no flow conditions. The temperature distribution inside the solar thermochemical reactor cavity subjected under no flow condition is comparatively more uniform and higher as compared to the nitrogen flow condition in both cases. Since, in case-2, the focal point is 20 mm inside the cavity thus it yields the higher temperature as compared to the case-1. This particular phenomenon occurs because the focal point of dish in case-1 is located at the cavity aperture but the flux does not quite gets distributed and covers the cavity walls enough to generate the higher temperature.

The temperature distribution in the radial direction in porous media is an important factor as it provides the information regarding the optimum catalyst (RPC) thickness. The results for radial temperature distribution in the porous RPC zone for case-1 and case-2 for flow, and no flow conditions has been shown in

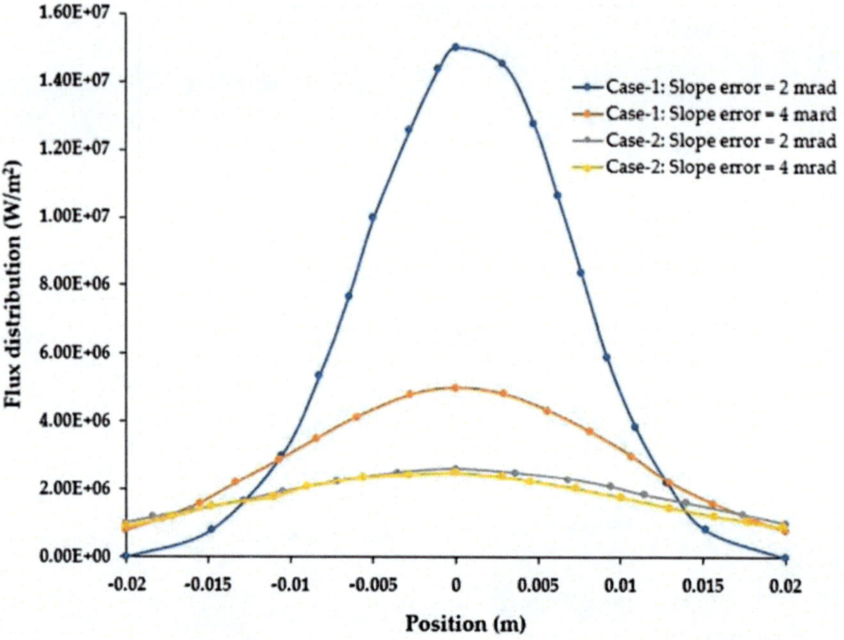


**Fig. 4** Flux distribution for case-1 and case-2

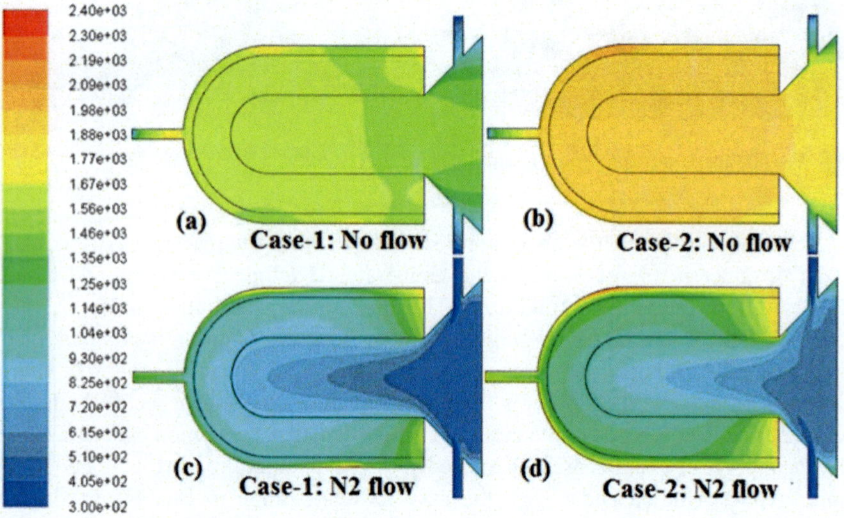


**Fig. 5** Temperature contours

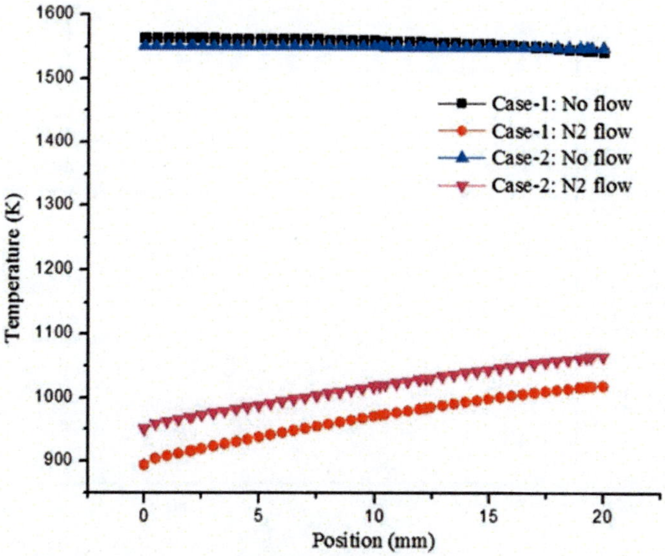


**Fig. 6** Radial temperature distribution in the RPC porous zone

Fig. 6. The temperature distribution along the radial direction in porous media under no flow condition for both cases is much higher due to the absence of cooling agent and achieve the almost same and uniform temperature above 1500 K. However, in flow condition, the temperature drops drastically but maintains the uniformity for both cases and yields the higher temperature for case-2.

Another important parameter of reactor cavity design analysis is the length of the porous medium; thus, it becomes necessary to analyze the temperature distribution along the axial direction in the porous media. Since the cylindrical part of the cavity has the length of 80 mm, the temperature distribution in axial direction is evaluated for this portion only and the dome part of the cavity is covered in the full reactor temperature analysis along the centerline. The axial temperature distribution has been shown in Fig. 7. The temperature for no flow condition in both cases is presumably higher compared to sweep gas flow condition and also remains uniform along the 80 mm length of the RPC porous zone. However, in nitrogen flow case, the temperature at ($x = 0$) cavity aperture is considerably high, and it drops from $x = 0$ to $x = 80$ mm due to the flux distribution, and difference in temperature for both cases is quite less.

To analyze the optimum temperature to carry out redox reaction, it is necessary to evaluate the reactor temperature. Figure 8 shows the fluid temperature distribution in the reactor cavity along the centerline. The temperature for no flow condition is uniform and remains above 1500 K and perfectly suited to carry out the redox reactions. It can be seen in the plot that the case-2 yields the higher reactor temperature compared to case-1 under nitrogen flow condition. At $x = 0$ (cavity aperture), the temperature is at over 400 K and it goes up to 1000 K at $x = 125$ mm.

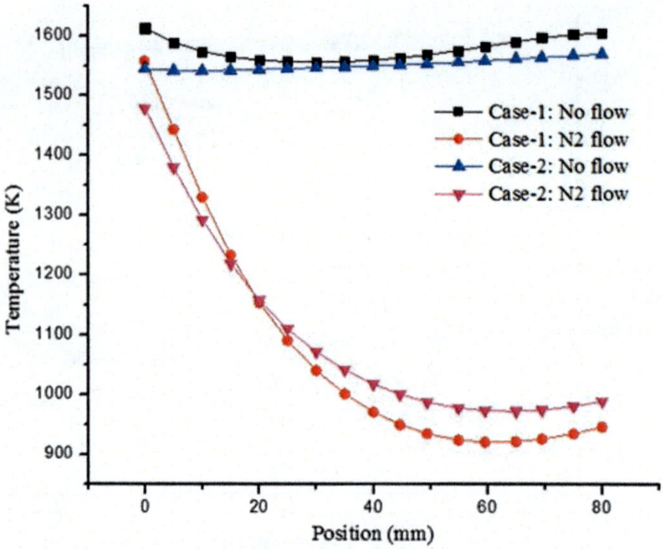


**Fig. 7** Axial temperature distribution in the RPC porous zone

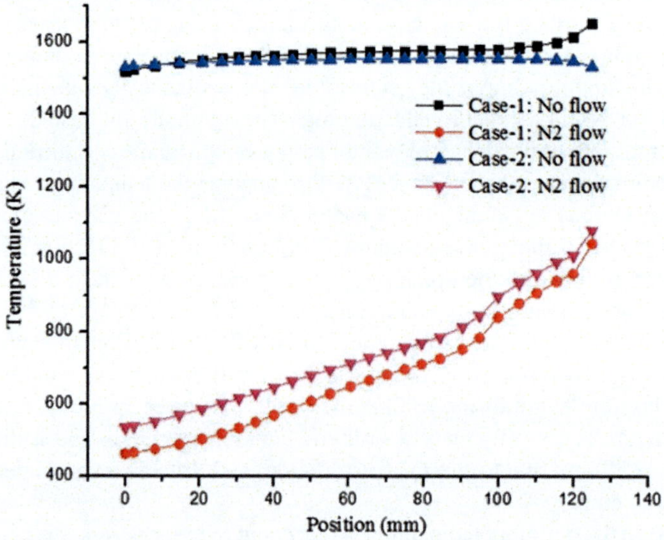


**Fig. 8** Fluid phase temperature distribution in the STCR cavity along the centerline

The flux profile is affected by the optical errors of concentrator. The optical error is mainly due to slope error which typically vary from 2 to 4 mrad. The variations in the solar flux profile due to slope errors can be seen in Fig. 4. It can be observed from the figure that the magnitude of the flux profile decreases as the slope error increases.

## 6 Conclusion

This study presented the effects of geometrical and optical parameters on incident solar flux distribution in solar thermochemical reactor cavity. Based on the result obtained from the numerical investigation, the following conclusions are drawn:

1. The location of focal point affects the magnitude of solar flux, and also, the distribution inside the solar thermochemical reactor cavity gets affected significantly.
2. When the focal point is at cavity aperture, it allows the incoming solar flux to uniformly distribute on the cavity walls and raise the temperature of catalyst region. However, when the focal point is inside the reactor cavity, the flux gets uniformly distributed but the magnitude of flux gets significantly reduced; thus, it does not attains the high temperature in the reactor cavity as compared to case-2.
3. The magnitude solar flux for case-2 with slope error 2 mrad is 50% less than case-1 with slope error 4 mrad and 16.67% less than case-1 with slope error 2 mrad.

The results given above provide geometrical and optical parameters as part of current research on solar thermochemical reactor cavity. The design of the reactor cavity and reducing the thermal losses (convective and radiative) of reactor cavity is major concern and needs to be investigated.

## References

1. Perkins C, Weimer AW (2004) Likely near-term solar-thermal water splitting technologies. Int J Hydrogen Energy 29(15):1587–1599
2. Agrafiotis C, von Storch H, Roeb M, Sattler C (2014) Solar thermal reforming of methane feedstocks for hydrogen and syngas production—a review. Renew Sustain Energy Rev 29:656–682
3. Hirsch D, Steinfeld A (2004) Solar hydrogen production by thermal decomposition of natural gas using a vortex-flow reactor. Int J Hydrogen Energy 29(1):47–55
4. Müller R, Zedtwitz Pv, Wokaun A, Steinfeld A (2003) Kinetic investigation on steam gasification of charcoal under direct high-flux irradiation. Chem Eng Sci 58(22):5111–5119
5. Montini T, Melchionna M, Monai M, Fornasiero P (2016) Fundamentals and catalytic applications of $CeO_2$-based materials. Chem Rev 116(10):5987–6041

6. Otake T, Yugami H, Yashiro K, Nigara Y, Kawada T, Mizusaki J (2003) Nonstoichiometry of Ce1-XYXO 2-0.5X-δ (X = 0.1, 0.2). Solid State Ionics 161(1–2):181–186
7. Jilte RD, Kedare SB, Nayak JK (2014) Investigation on convective heat losses from solar cavities under wind conditions. In: Energy procedia, vol 57
8. Jilte RD, Nayak JK, Kedare SB (2017) Experimental investigation on heat losses from differentially heated cylindrical cavity receiver used in paraboloid concentrator. J Sol Energy Eng 139(3):
9. Jilte RD, Kedare SB, Nayak JK (2013) Natural convection and radiation heat loss from open cavities of different shapes and sizes used with dish concentrator. Mech Eng Res 3(1):25
10. Loni R, Kasaeian AB, Askari Asli-Ardeh E, Ghobadian B (2016) Optimizing the efficiency of a solar receiver with tubular cylindrical cavity for a solar-powered organic Rankine cycle. Energy 112:1259–1272
11. Harris JA, Lenz TG (1985) Thermal performance of solar concentrator/cavity receiver systems. Sol Energy 34(2):135–142
12. Delatorre J et al (2014) Monte Carlo advances and concentrated solar applications. Sol Energy 103:653–681
13. Li H, Huang W, Huang F, Hu P, Chen Z (2013) Optical analysis and optimization of parabolic dish solar concentrator with a cavity receiver. Sol Energy 92:288–297
14. Daabo AM, Mahmoud S, Al-Dadah RK (2016) The effect of receiver geometry on the optical performance of a small-scale solar cavity receiver for parabolic dish applications. Energy 114:513–525
15. Daabo AM, Mahmoud S, Al-Dadah RK (2016) The optical efficiency of three different geometries of a small scale cavity receiver for concentrated solar applications. Appl Energy 179:1081–1096
16. Daabo AM, Al-Mola YS, Al-Rawy AY, Lattimore T (2019) State of the art single-objective optimization of small scale cylindrical cavity receiver. Sustain Energy Technol Assess 35 (June):278–290
17. Daabo AM, Mahmoud S, Al-Dadah RK, Ahmad A (2017) Numerical investigation of pitch value on thermal performance of solar receiver for solar powered Brayton cycle application. Energy 119:523–539
18. Pavlovic S, Daabo AM, Bellos E, Stefanovic V, Mahmoud S, Al-dadah RK (2017) Experimental and numerical investigation on the optical and thermal performance of solar parabolic dish and corrugated spiral cavity receiver. J Clean Prod 150:75–92
19. Wendelin T (2003) Soltrace: a new optical modeling tool for concentrating solar optics. In: International solar energy conference
20. Cooper T, Steinfeld A (2011) Derivation of the angular dispersion error distribution of mirror surfaces for monte carlo ray-tracing applications. J Sol Energy Eng 133(4)
21. Luo Y, Du XZ, Yang LJ, Yang YP (2014) Numerical simulation on the performance of a combination of external and cavity absorber for solar power plant. Energy Proc 49:428–437
22. Binotti M, Di Marcoberardino G, Biassoni M, Manzolini G (2017) Solar hydrogen production with cerium oxides thermochemical cycle. In: AIP Conf Proc, vol 1850
23. Wendelin T, Dobos A, Lewandowski A, Wendelin T, Dobos A (2013) SolTrace : a ray-tracing code for complex solar optical systems soltrace : a ray-tracing code for complex solar optical systems
24. Lee H (2014) The geometric-optics relation between surface slope error and reflected ray error in solar concentrators. Sol Energy 101:299–307

# Query Optimization of Corporate Service Quality Using Log Shipping Architecture

**Sunita Rani, Bhupendra Kumar, Deepkiran Munjal, and Shweta Chaudhary**

**Abstract** The current research is an empirical assessment of measuring the effectiveness of the application of query optimizing log shipping architecture in corporate India. Theoretically stating, the major aim of integrating such database architecture at a workplace aims at reducing the technical debt and number of fixes required along with increasing the efficiency of maintenance of transaction logs and data retrieval with avoidance of any kind of failure. This research study has been conducted in order to verify the efficiency of the working of these optimizers in the corporate world. Thus, the major objective of this research is to measure the effectiveness of the stipulated database architecture through various variables of service quality including reliability, responsiveness, security, competence and durability. The study further aims at examining the impact of these variables on the effectiveness of the stipulated architecture. The sample included the technical staff working in the corporate sector in the technical or IT department in several companies located in Delhi-NCR based on convenience sampling. The data were collected through a self-structured questionnaire that was floated to online to the respondents comprising a total sample of 400 respondents. The questionnaire was reliable as the Cronbach's alpha was greater than 0.75. Its validity was also checked by some professionals in this field along with some academicians working for the same. Multiple regression analysis has been used to examine the impact of the given variables on the effectiveness of the query optimizing log shipping architecture at the workplace. The results provided insights regarding the level of effectiveness of the database architecture in the organizations in order to gain efficiency. This study has been quite relevant as it empirically tested the

S. Rani
Jagannath International Management School, New Delhi, India

B. Kumar (✉)
Mangalmay Institute of Engineering and Technology, Greater Noida, India
e-mail: Bhupendra.bhadana@mietengineering.org

D. Munjal
GL Bajaj Institute of Technology, Greater Noida, India

S. Chaudhary
ABES Engineering College, Ghaziabad, India

 
R. Kumar et al. (eds.), *Recent Trends in Thermal Engineering*, Lecture Notes in Mechanical Engineering, https://doi.org/10.1007/978-981-16-3132-0_23

effectiveness of the given query optimizing architecture along with identifying the factors impacting the effectiveness. Also, it will add as a valuable piece of knowledge to the existing literature.

**Keywords** Log shipping architecture · Optimization · Transaction log · Service quality

## 1 Introduction

This paper aims at studying the effectiveness of the query optimizing log shipping architecture in terms of service quality. Before moving to the analytical part, it is important to first understand the significance and relevance of the same. It is documented that logging architecture is indeed an important part concerning the application health as well as the performance monitoring with respect to the modern distributed application architecture world. In this massively developing corporate world, the fast paced and hyper-scale deployment along with efficient automation absolutely depends on the logging information in order to identify the status or operational behaviour of stipulated applications, networks and infrastructure [1]. The providers of these architectures are thoroughly involved in advancements and create a constant push for others to keep innovating to provide better architectures due to which there is constant competition in this industry [2].

It is to be understood that once log shipping has been configured, the information regarding the status of all the log shipping servers can be monitored. Also, the history and status of log shipping operations are always saved locally by the log shipping jobs. Further, the history along with the status of the backup operations gets stored at the primary server followed by the storing of the history and status of the copy and restore operations take place in the secondary server [3]. Query optimization is used to optimize the efficiency results of operations by achieving the targets of less time and low costs [4]. Hence, undoubtedly, the query optimizing log shipping structure is extremely important in the current technology-driven work environment. Thus, it becomes quite important to examine the viewpoint of the users of this technical adaptation by the corporate world.

Efficiency in terms of wealth maximization is what every organization aspires [5]. This can be simply achieved through cost reduction and getting the work done in time with minimum or no lag time, in order to measure the effectiveness of the stipulated architecture in the database of the organizations making the corporate India. It is measured through the variables of service quality comprising reliability, responsiveness, security, competence and durability in this empirical research. Service quality basically reflects whether the expectations or perceived performance is matching the actual performance [6]. Service quality in this study reflects the effectiveness in terms of the performance of the query optimizing log shipping architecture in the corporate offices in Delhi-NCR.

## 2 Literature Review

This section of the paper deals with reviewing the studies conducted by researchers pertaining to the query optimizing log shipping architectures in order to identify the research gaps that would justify the conducting of the current study. Gupta and Kashinath [7] conducted a research to study the query processing and optimization in the distributed architecture using NoSql with respect to a banking system or sector. It documented the structures of operations of the query optimizers in the bank branches. Also, Mehta [4] presented a detailed review model on query optimization using the open-source database. In this study, the different approaches towards query optimization were reviewed. The various kinds of approaches that were discussed included constraint-based pruning approach, semantic-based query optimization and cost-based optimization technique. Samanth et al. [8] made a documentation about the query optimization issues for data retrieval in the case of cloud computing. These issues were pertaining to searching techniques and the time duration when a query is optimized. Looy and Shafagatova [9] studied the business process performance measurement through a structured literature review of indicators, measures and metrics. This study aimed at studying in detail the different aspects and approaches adopted by businesses for their performance measurement. From the given review of the past studies, the research gaps that emerge reflect the requirement of measuring effectiveness of query optimizers across different businesses and not just a particular sector. Also, secondary research in this regard is ample yet, there is sufficient scope for empirical assessment towards the evaluation of the effectiveness of such architectures in database in the organizations. Also, this branch of study requires consistent review over definite periods of time due to which it is further quite relevant.

Service quality concept has been reviving from time to time. There have been several studies conducted in different sectors to measure the service quality from consumers in different streams. Raza et al. [10] presented an analysis that showed a significant positive relationship between the factors including reliability, tangibility, assurance and responsiveness with the customer satisfaction. Sakhaei et al. [11] also highlighted the variable, reliability as a significant dimension of service quality. Ariff et al. [12] had also highlighted that competence reflected through the factors such as efficiency, fulfilment, proper system availability, website aesthetic value and the proper guidance given were really crucial for ensuring the quality of electronic service quality that would generate satisfaction and loyalty. Gupta and Bansal [13] also postulated that there were indeed several other quality dimensions, out of which reliability was a prominent dimension that determined the level of customer satisfaction. There were various other dimensions which comprised security, responsiveness, efficiency and website aesthetics which were also important in the determination of the customer satisfaction. Firdous and Farooqi [14] conducted an exploratory survey in order to study the various dimensions of service quality and further understand its impact on customer satisfaction. Having reviewed the variables from the past studies, suitable variables have been selected considering the current study.

It was also found that extensive research had taken place to measure service quality for assessing the satisfaction level, performance, effectiveness and efficiency across various industries. The respondents have been the users of the chosen products or services. For instance, Kumar and Shenbagaraman [15] undertook a primary study in order to investigate the customer's perception towards online banking and online service quality along with its influence on the customer satisfaction in Chennai. Tharanikaran et al. [16] aimed at assessing the degree of service quality along with customer satisfaction in the electronic banking domain via this study in Battticaloa district—Akkucuk and Teuman [17] conducted an extensive review of literature on service quality and e-service quality. Velmurugan [18] carried an empirical study with the objective of ascertaining the customer perception towards service quality of public sector banks. Bharwana et al. [19] carried an empirical assessment regarding the assessment of service quality on customer satisfaction with respect to the private colleges in Faisalabad, Punjab in Pakistan. Apart from this, there have been several other studies on service quality in other sectors, but somehow, sufficient literature does not exist to measure effectiveness of performance of the query optimizers in the technical domain in corporate India. Therefore, the current study is justified as it explores the same in the stipulated domain.

The various independent variables identified for this study have been reliability, responsiveness, security, competence and durability. Further, the specific elements for each variable studied through this research are shown in Figs. 1, 2, 3, 4 and 5, respectively.

**Fig. 1** Elements of reliability

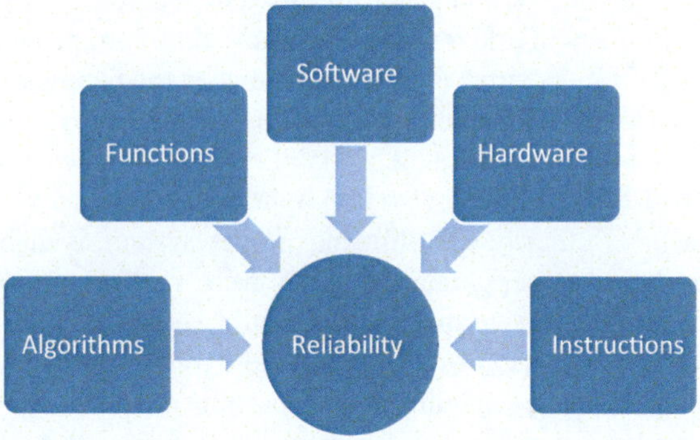

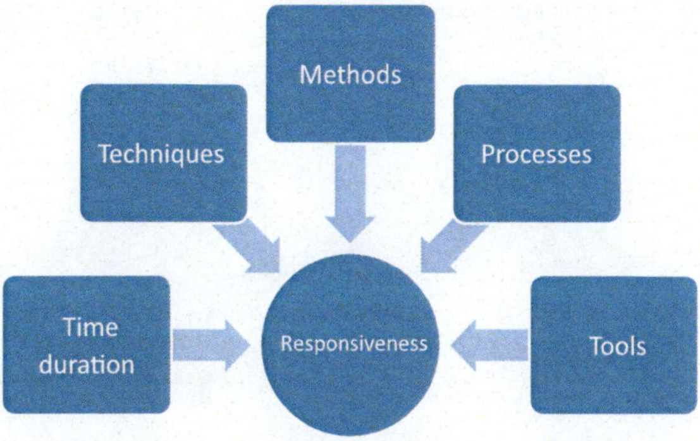


**Fig. 2** Elements of responsiveness

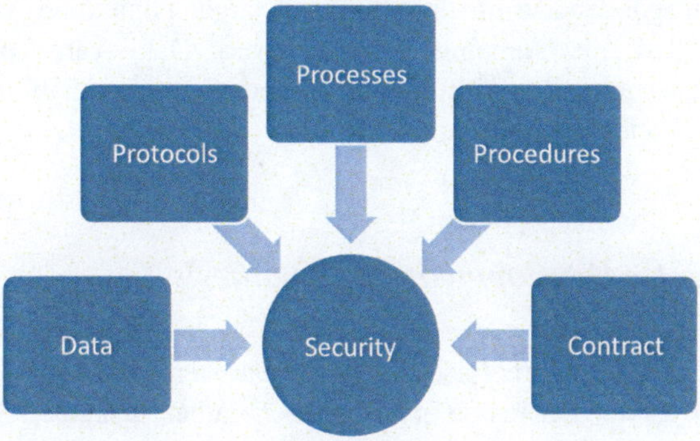


**Fig. 3** Elements of security

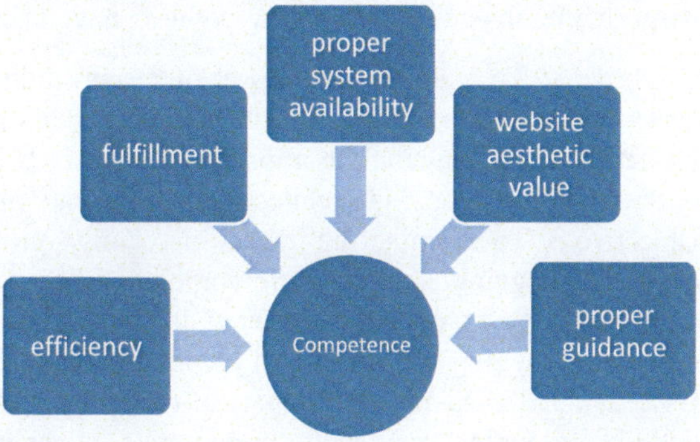


**Fig. 4** Elements of competence

# 3 Research Methodology

This study followed a descriptive research design. The target population included the technical professionals or staff working with query optimization log shipping architecture or database in the corporate business houses located in Delhi-NCR. Convenience sampling was used for choosing the sample that constituted 400

Fig. 5 Elements of durability

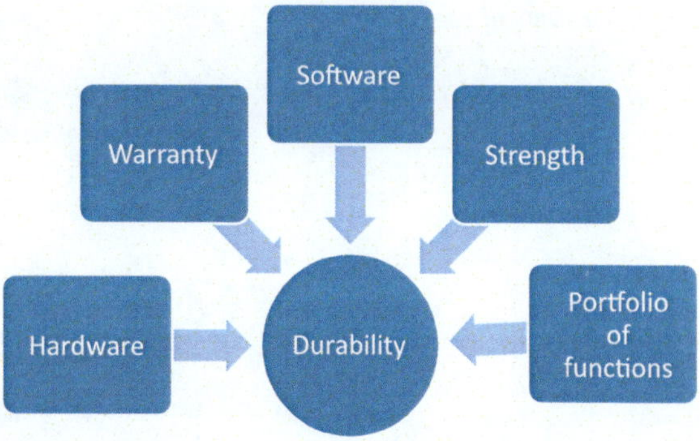


respondents from the stipulated target population, out of which 398 completely filled questionnaires were received. Data were collected via a self-structured questionnaire with major components as reliability, responsiveness, security, competence and durability.

## *3.1 Development of Hypothesis*

Five statements for each component were constructed on a Likert's five-point scale with the scores as 1-disagree, 2-rather disagree, 3-neutral, 4-rather agree and 5-agree. It is to be noted that the summation of scores for all the variables reflected the effectiveness of the query optimization log shipping architecture. For this purpose, the objectives and null hypothesis have been stated below:

Objective 1: To examine the impact of responsibility on the effectiveness of the query optimizing logging architecture in corporate offices in Delhi-NCR.
Objective 2: To examine the impact of responsiveness on the effectiveness of the query optimizing logging architecture in corporate offices in Delhi-NCR.
Objective 3: To examine the impact of security on the effectiveness of the query optimizing logging architecture in corporate offices in Delhi-NCR.
Objective 4: To examine the impact of competence on the effectiveness of the query optimizing logging architecture in corporate offices in Delhi-NCR.
Objective 5: To examine the impact of durability on the effectiveness of the query optimizing logging architecture in corporate offices in Delhi-NCR.
H01: No significant impact of responsibility on the effectiveness of the query optimizing logging architecture in corporate offices in Delhi-NCR.
H02: No significant impact of responsiveness on the effectiveness of the query optimizing logging architecture in corporate offices in Delhi-NCR.
H03: No significant impact of security on the effectiveness of the query optimizing logging architecture in corporate offices in Delhi-NCR.

H04: No significant impact of competence on the effectiveness of the query optimizing logging architecture in corporate offices in Delhi-NCR.
H05: No significant impact of durability on the effectiveness of the query optimizing logging architecture in corporate offices in Delhi-NCR.

Multiple regression was applied to examine the relationships between the variables and test the null hypothesis. The conceptual framework for measuring the effectiveness of the query optimizing logging architecture is shown in Fig. 6.

## 3.2 *Model Development and Validation*

This section comprises the representation of data after the application of the research techniques to obtain meaningful interpretation of the collected data that were first coded, tabulated and analysed using a research tool.

It is documented in Table 1 that the mean score for effectiveness for the current study was 103.50 (S.d. = 42.075) with highest mean score for reliability being 2.98 (S.d. = 1.277) and the lowest mean score for competence as 2.95 (S.d. = 1.314). The total number of respondents as represented by $N$ was 398.

Further, Table 2 has shown statistically positive correlation between the mentioned independent variables including reliability, responsiveness, security, competence, durability and effectiveness of the query optimizing ship logging architecture as $p < 0.05$.

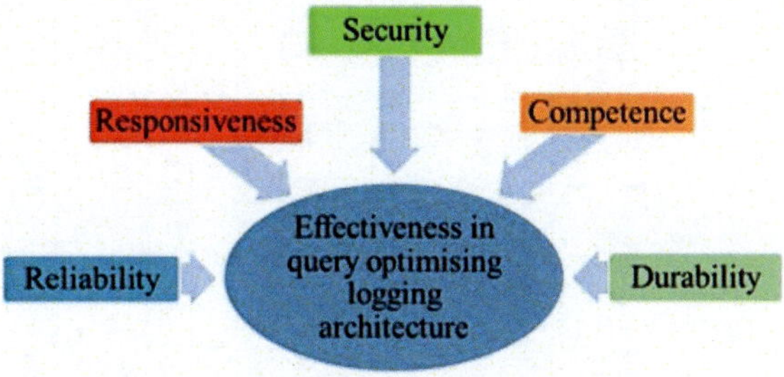


**Fig. 6** Conceptual framework

**Table 1** Descriptive statistics for effectiveness and service quality variables

| | Mean | Std. deviation | $N$ |
|---|---|---|---|
| Effectiveness | 103.50 | 42.075 | 398 |
| Reliability | 2.98 | 1.277 | 398 |
| Responsiveness | 2.96 | 1.315 | 398 |
| Security | 2.96 | 1.308 | 398 |
| Competence | 2.95 | 1.314 | 398 |
| Durability | 2.96 | 1.332 | 398 |

**Table 2** Correlation matrix for effectiveness and service quality variables

| | | Effectiveness | Reliability | Responsiveness | Security | Competence | Durability |
|---|---|---|---|---|---|---|---|
| Pearson correlation | Effectiveness | 1.000 | 0.899 | 0.918 | 0.923 | 0.906 | 0.895 |
| | Reliability | 0.899 | 1.000 | 0.817 | 0.800 | 0.778 | 0.794 |
| | Responsiveness | 0.918 | 0.817 | 1.000 | 0.831 | 0.795 | 0.781 |
| | Security | 0.923 | 0.800 | 0.831 | 1.000 | 0.850 | 0.793 |
| | Competence | 0.906 | 0.778 | 0.795 | 0.850 | 1.000 | 0.794 |
| | Durability | 0.895 | 0.794 | 0.781 | 0.793 | 0.794 | 1.000 |
| Sig. (1-tailed) | Effectiveness | 0.000 | 0.001 | 0.001 | 0.001 | 0.001 | 0.001 |
| | Reliability | 0.001 | 0.000 | 0.001 | 0.001 | 0.001 | 0.001 |
| | Responsiveness | 0.001 | 0.001 | 0.000 | 0.001 | 0.001 | 0.001 |
| | Security | 0.001 | 0.001 | 0.001 | 0.000 | 0.001 | 0.001 |
| | Competence | 0.001 | 0.001 | 0.001 | 0.001 | 0.000 | 0.001 |
| | Durability | 0.001 | 0.001 | 0.001 | 0.001 | 0.001 | 0.000 |
| N | Effectiveness | 398 | 398 | 398 | 398 | 398 | 398 |
| | Reliability | 398 | 398 | 398 | 398 | 398 | 398 |
| | Responsiveness | 398 | 398 | 398 | 398 | 398 | 398 |
| | Security | 398 | 398 | 398 | 398 | 398 | 398 |
| | Competence | 398 | 398 | 398 | 398 | 398 | 398 |
| | Durability | 398 | 398 | 398 | 398 | 398 | 398 |

**Table 3** Corresponding beta values and significance levels of each variable

| Variables | Beta | Significance |
|---|---|---|
| Reliability | 0.190 | $p < 0.05$ |
| Responsiveness | 0.253 | $p < 0.05$ |
| Security | 0.227 | $p < 0.05$ |
| Competence | 0.199 | $p < 0.05$ |
| Durability | 0.209 | $p < 0.05$ |

Enter method was used for the application of multiple regression. Table 3 consists of the corresponding beta values and significance levels of each variable. The developed linear regression model is shown in Eq. 1.

From the above details, the following regression equation has been obtained:

$$\begin{aligned}\text{Effectiveness} = {} & 0.190(\text{Reliability}) + 0.253(\text{Responsiveness}) + 0.227(\text{Security}) \\ & + 0.199(\text{Competence}) + 0.209(\text{Durability})\end{aligned} \quad (1)$$

The values of R2 (coefficient of determination) and R2-adjusted (adjusted coefficient of determination) have been observed as 0.990 and 0.979, respectively. These values show that developed model has higher degree of fitness.

Analysis of variance (ANOVA) has been performed to validate the adequacy of developed regression model and shown in Table 4. In ANOVA, $F(5392) = 3728.63$ at $p < 0.05$, which showed the model has statistically significant value at 5% level of confidence.

Predictors: constant, durability, responsiveness, competence, reliability and security.

This implies that with 0.190 units change in reliability, the effectiveness for query optimizing ship logging architecture changes by one unit. Since, $p < 0.05$, thus the impact is statistically significant at 5% level of significance. Hence, H01 is rejected. It implies that there was a significant impact of reliability on the query optimizing log shipping architecture in the corporate offices in Delhi-NCR. Similarly, with 0.253 units change in responsiveness, the effectiveness changes by one unit. Since, $p < 0.05$, thus the impact is statistically significant at 5% level of significance. Hence, H02 is rejected. It implies that there was a significant impact of responsiveness on the query optimizing log shipping architecture in the corporate offices in Delhi-NCR. Further, with 0.227 units change in security, the effectiveness changes by one unit. Since, $p < 0.05$, thus the impact is statistically significant at 5% level of significance. Hence, H03 is rejected. It implies that there was a significant impact of security on the query optimizing log shipping architecture in the corporate offices in Delhi-NCR. Also, with 0.199 units change in competence, the effectiveness changes by one unit. Since, $p < 0.05$, thus the impact is statistically significant at 5% level of significance. Hence, H04 is rejected. It implies that there was a significant impact of competence on the query optimizing log shipping architecture in the corporate offices in Delhi-NCR. With 0.209 units change in

**Table 4** ANOVA table

| Model | Sum of squares | Df | Mean square | *F* | Sig. |
|---|---|---|---|---|---|
| Regression | 688349.94 | 5 | 137669.99 | 3728.63 | <0.001 |
| Residual | 14473.54 | 392 | 36.92 | | |
| Total | 702823.48 | 397 | | | |

Dependent variable: effectiveness
Predictors: constant, durability, responsiveness, competence, reliability, security

**Table 5** Coefficients matrix

| Model | Unstandardized coefficients | | Standardize coefficients | T | Sig. |
|---|---|---|---|---|---|
| | B | Std. error | Beta | | |
| (Constant) | 0.987 | 0.814 | | 1.213 | 0.226 |
| Reliability | 6.254 | 0.477 | 0.190 | 13.123 | 0.001 |
| Responsiveness | 8.092 | 0.483 | 0.253 | 16.754 | 0.001 |
| Security | 7.293 | 0.525 | 0.227 | 13.881 | 0.001 |
| Competence | 6.356 | 0.487 | 0.199 | 13.040 | 0.001 |
| Durability | 6.611 | 0.441 | 0.209 | 14.992 | 0.001 |

Dependent variable: effectiveness

durability, the effectiveness changes by one unit. Since, $p < 0.05$, thus the impact is statistically significant at 5% level of significance. Hence, H05 is rejected. It implies that there was a significant impact of durability on the query optimizing log shipping architecture in the corporate offices in Delhi-NCR. Table 5 shows the coefficient matrix of all model variables.

It has been found from this empirical assessment that all the null hypothesis H01, H02, H03, H04 and H05 were rejected. This implies that all the independent variables comprising reliability, responsiveness, security, competence and durability impacted the effectiveness of query optimizing ship logging architecture at 5% level of significance as per the scores obtained from the primary survey conducted. Hence, the functioning of these database architectures has been found to be dependent upon the stipulated variables in order to produce optimized results by bring down technical flaws and generating efficient fixes. This further helps the organizations work efficiently with the attainment of the objectives of cost reduction and completing the assigned tasks in optimal time duration. All this further contributes to wealth maximization.

## 4 Conclusions

It is concluded from this study that the users of the query optimizing ship logging architecture in corporate India documented the performance of these architectures being governed by elements of reliability, responsiveness, security, competence and durability. The overall score has been above average for the effectiveness of performance, and also, all the chosen variables displayed a statistically significant impact on the effectiveness of the query optimizers. Undoubtedly, it is an input for the companies dedicated towards constructing and providing these architectures to the corporate world to direct efforts for the enhancement of performance based on the proposed variables.

## References

1. Sabnis S (2018) Centralized and externalized logging architecture for modern applications using rack scale flash storage. Database zone. Retrieved from: https://dzone.com/articles/centralized-and-externalized-logging-architecture
2. Jorgensen A et al (2014) Professional Microsoft SQL server 2014 administration. Wiley, United States
3. Mighty Pen et al (2017) Monitor log shipping (transact-SQL). Retrieved from: https://docs.microsoft.com/en-us/sql/database-engine/log-shipping/monitor-log-shipping-transact-sql?view=sql-server-ver15
4. Mehta M (2013) A review model on query optimization using open source database. Int J Sci Eng Res 4(4):1369–1373
5. Jones TM, Felps W (2013) Shareholder wealth maximization and social welfare: a utilitarian critique. Bus Ethics Q 23(2):207–238. Retrieved from: https://doi.org/10.5840/beq201323215
6. Borgave S, Koranne S (2012) Service quality management: a literature review. Ethos 5(2): 1–24. Retrieved from: https://www.researchgate.net/publication/274332133_Service_Quality_Management_A_Literature_Review
7. Gupta P, Kashinath KC (2014) Query processing and optimization in the distributed architecture using NoSql: a banking system. Int J Adv Comput Eng Commun Technol 3(4):25–28. Retrieved from: www.irdindia.in/journal_ijacect/pdf/vol3_iss4/5.pdf
8. Samanth N et al (2012) Query optimization issues for data retrieval in cloud computing. Int J Comput Eng Res 2(5):1361–1364
9. Looy AV, Shafagatova A (2016) Business process performance measurement: a structured literature review of indicators, measures and metrics. Springer Plus 5(1):1–41
10. Raza et al (2015) Internet banking and customer satisfaction in Pakistan. Qual Res Financ Markets 7(1):24–36
11. Sakhaei et al (2013) The impact of service quality on customer satisfaction in internet banking. J Math Comput Sci 99(1):33–40
12. Ariff et al (2014) The impacts of E-service quality and E-customer satisfaction on E-customer loyalty in internet banking. Adv Sci Lett 20(1):285–289
13. Gupta KK, Bansal DI (2012) Development of an instrument to measure internet banking service quality in India. Int Referred Res J 3(2):11–25
14. Firdous S, Farooqi R (2017) Impact of internet banking service quality on customer satisfaction. J Internet Bank Commer 22(1):117

15. Kumar G, Shenbagaraman VM (2017) A study on customer's perception of online banking and E-service quality among Chennai customers. Int J Bus Excellence 11(1):72–94. Retrieved from: https://doi.org/10.1504/IJBEX.2017.080606
16. Tharanikaran V et al (2017) Service quality and customer satisfaction in the electronic banking. Int J Bus Manag 12(4):67–83
17. Akkucuk U, Teuman B (2016) Assessing service quality in online banking services. Prob Perspect Manag 14(2–1):182–192
18. Velmurugan R (2015) Service quality of public sector banks. Indian J Res 4(4):3–6
19. Bharwana TK et al (2013) Impact of service quality on customers' satisfaction: a study from service sector especially Private Colleges of Faisalabad, Punjab, Pakistan. Int J Sci Res Publ 3 (5):1–7. Retrieved from: http://www.ijsrp.org/research-paper-0513/ijsrp-p1768.pdf

# Wear Characterisation of Stir Cast Mg-SiC Composites

**Avtar Singh, Niraj Bala, and Munish Mehta**

**Abstract** Composites (metal matrix) are known to be ideal materials for improving the properties of components used in automotive and aviation industries. This research work deals with magnesium matrix composite produced by stir casting method by using two weight percentages (4 and 7 wt.%) of silicon carbide reinforcement material. Distribution of reinforcement particles in matrix material is studied by optical microscope and electron microscopy scanning techniques. Hardness as well as wear behaviour of the composites has been examined and compared with pure magnesium. Composite produced with 4 and 7 wt.% showed improved hardness around 26.67% and 35.56%, respectively. Pin on disc wear tests were performed by varying loads (5 to 30 N). Silicon carbide-reinforced composite (7 wt.%) showed more wear resistance than the other composite and pure magnesium. Ploughing marks and oxide surfaces were found in all worn surfaces of wear specimens; however, ploughing marks were found to be deeper in worn surface of pure magnesium at 30 N. Lot of debris and agglomeration of debris have been found on the worn surface of as cast pure Mg as equated to Mg-SiC composites.

**Keywords** Magnesium matrix composite · SiC · Hardness · Wear · SEM

## 1 Introduction

In the present scenario, the automobile industry is paying all attention for the development of metal matrix composites because as compared to conventional materials, these materials are light in weight [1]. Magnesium matrix-based

A. Singh (✉)
IKGPTU, Kapurthala, India

N. Bala
Department of Entrepreneurship Development & Industrial Coordination, NITTTR, Chandigarh, India

M. Mehta
School of Mechanical Engineering, Lovely Professional University, Phagwara, India

 
R. Kumar et al. (eds.), *Recent Trends in Thermal Engineering*, Lecture Notes in Mechanical Engineering, https://doi.org/10.1007/978-981-16-3132-0_24

composites are attaining value as good material in the automobile sector to reduce pollution [2]. Density of magnesium is about 66.67% of aluminium, 25% of zinc and 20% of steel [3]. Addition of reinforcement materials to magnesium alloys may enhance material properties, like strength, hardness and wear resistance [4, 5]. Various processes are used to enhance properties of matrix material, and addition of reinforcements to the matrix material by stir casting is amongst the most useful methods [6]. Poddar et al. [5] analysed that the existence of SiC particles has led to a significant enhancement in the hardness, yield strength of the AZ91D/SiC composite produced by the stir casting process. Hardness and yield strength increased as reinforcement content increased. Composites manufactured by semi-solid stirring method by using three weight percentages (3, 6 and 9) of SiC show uniformly distributed particles in the AZ81A matrix [7]. Aravindan et al. [8] manufactured AZ91D composites reinforced by varying vol. % of SiC particles by using stir casting method. Study showed that the composite properties increased with an increase in SiC particles and decreased with an increase in particle size. Viswanath et al. [9] produced AZ91 alloy-based composite reinforced with different wt.% (5 to 25%) of SiC by using stir casting method. The microstructure study showed the uniform distribution of particles with very less agglomerations. Kandpal et al. [10] fabricated a composite using stir casting methods and showed improved micro-hardness and ultimate tensile strength. Cubic boron nitride reinforced with SiC whisker developed by sintering showed better mechanical properties such as hardness and fracture strength in comparison with the unreinforced material [11].

The main intention of current research work is to examine the wear behaviour of composites produced by using stir casting method. Two wt.% (4 and 7 wt.%) of reinforcement were chosen to fabricate the composites. A similar method has also been used to produce a monolithic magnesium sample for comparative analysis.

## 2 Experimental Methodology

### *2.1 Composites Fabrication*

Pure magnesium was used as matrix material in present study. SiC has been used as reinforced material with 4 and 7 wt.% to fabricate composites. Pure Mg ingots have been cut into small pieces for stir casting process. Castings were prepared by using method explained somewhere else by Singh and Bala [12]. Specimens of composites (Mg-SiC) and pure magnesium were prepared for the test of hardness and wear from casting blocks. The designation of the specimens used in this paper is as follows: pure magnesium as pure Mg, composite with 4 wt.% of SiC is Mg-4SiC and with 7 wt.% is Mg-7SiC.

## 2.2 *Microstructure Analysis*

To study the distribution of particles of reinforcement, the microstructure analysis was done. For this analysis, the 5-mm-thick plate was cut from each casting with the help of electric discharge machine. These plates were further cut to square cross section of 20 × 20 mm. The polishing of the specimens was carried out using a regular technique followed by etching in solution. An optical microscope was used to examine microstructure of composites. Further structure analysis was done by using scanning electron microscope (SEM). The specimens were studied by SEM to examine the wear behaviour of the specimens.

## 2.3 *Hardness Test*

The Rockwell hardness testing machine is used to calculate the hardness. The Rockwell hardness test was conducted in order to examine the effect of reinforcement on matrix hardness. Five hardness values have been taken for each specimen, and average of these values has been stated in this manuscript.

## 2.4 *Wear Test*

A pin on disc machine, fabricated by Ducom wear Instruments, was used to conduct wear test. Square cross sectional (5 × 5 mm) specimen pins were prepared from casting of composites and pure Mg. These pins were used to be worn against a revolving steel alloy disc. Disc speed of 477 rpm and 0.04 m track diameter were used to conduct wear test. Before carrying out the wear test, specimens were polished by emery paper having mesh size up to 1200. For each specimen, four loads, i.e. 5 N, 10 N, 20 N and 30 N, were used at the sliding speed of 1 $ms^{-1}$. Test of wear was conducted for a sliding distance of 1800 m. An electronic balance machine with a precision of 0.001 g is used to calculate the mass of the specimen pre- and post-wear test to determine the weight loss (grams) for every specimen. SEM of wear surfaces was examined to study the wear mechanism.

# 3 Results and Discussion

## 3.1 *Microstructure Examination*

The composites produced in accordance with the procedure referred in Sect. 2.1 were analysed by means of optical microscope and by SEM examination.

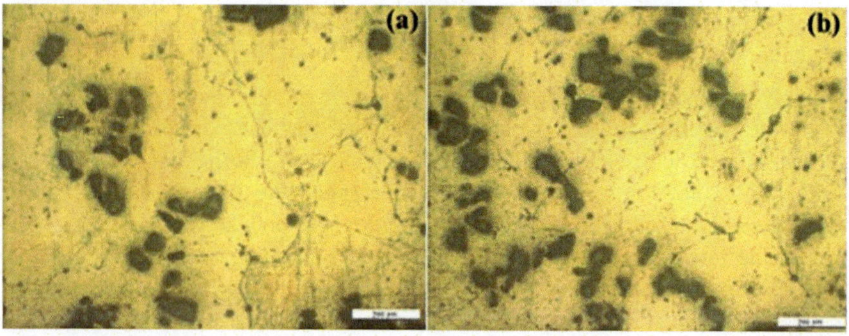


**Fig. 1** Optical images **a** Mg-4SiC and **b** Mg-7SiC

The optical microstructure images of Mg-SiC composites have been shown in Fig. 1a, b. Figure 1 clearly indicates microstructure along with grain boundaries. Number of SiC particles increased as the wt.% increased to 7 wt.% as shown in Fig. 1b. Some particles are lying on grain boundaries. The SEM images of composites Mg-4SiC (Fig. 2a) and Mg-7SiC (Fig. 2b) indicate that the reinforcement particles are almost uniformly spread as well as bonded well with the matrix. Study of AZ91/SiCp composite-based magnesium alloy has shown similar results that SiC particles are uniformly dispersed with far less accumulation of particles [13]. Mg/SiC composite fabricated by Gui et al. [14] showed that the SiC particles are uniformly dispersed and bound well to the matrix material.

### 3.2 Hardness Results

Hardness values of specimens have been shown in Table 1. Hardness results showed the enhancement in the hardness of composites. The results showed an increase in the hardness of the composites due to the presence of reinforcement, which avoids deformation of matrix material. The presence of reinforcements in matrix material enhances load-bearing ability and thereby increases the hardness [15]. Presence of SiC has increased composite hardness due to the interactive effect of the SiC phase that restricts focused matrix deformation [8]. Similar study showed that the hardness of composites based on Mg and AZ31 alloy matrix reinforced up to 10 wt.% of SiC got increased [16].

### 3.3 Wear Results

Wear is known to be a major issue in manufacturing applications. Ceramic reinforcements are being found to enhance the wear properties of Mg alloys [17].

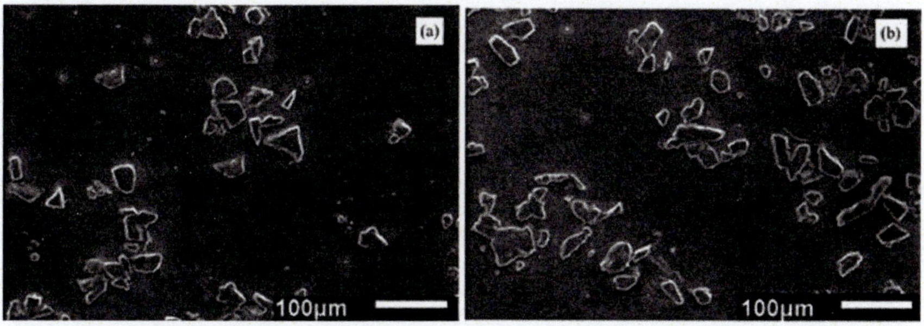


**Fig. 2** SEM images **a** Mg-4SiC and **b** Mg-7SiC

**Table 1** Hardness (HR) of pure Mg and Mg-SiC composites

| S. No. | Specimen designation | Hardness (HR) | Improvement in hardness (%) |
|---|---|---|---|
| 1 | Pure Mg | 45 | – |
| 2 | Mg-4SiC | 57 | 26.67% |
| 3 | Mg-7SiC | 61 | 35.56% |

Wear analysis was carried out by taking into account the material loss (in grammes) which occurred throughout the wear. As shown in Fig. 3a, the mass loss increased for all specimens with load increased. There is lesser contact area between the specimen and the steel disc at a lower load, which cause lesser mass loss of the specimens. With increased load, there was also increase in contact area, leading to an increase in weight loss of specimens [18]. The Mg-SiC composites show enhanced wear resistance as equated to pure magnesium. Similar results were observed by one of the researchers that the Mg/SiC composite showed higher wear resistance than the magnesium measured at loads varying from 5 to 50 N [1]. Although Mg-7SiC showed more wear resistance as compared to other specimens. As explained in Sect. 3.2, also Mg-7SiC showed higher hardness. Reinforcement particles can serve as a blockade and prevent plastic deformation, resulting in substantial decrease in mass loss [19]. There was a reduction in wear mass loss in composites relative to pure Mg. There is a major difference between pure Mg wear mass loss and composites as shown in Fig. 3b. As the reinforcement wt.% increased, the wear mass loss reduced for all load conditions.

### *3.4 Wear Surface Analysis*

SEM study examines the worn surfaces of specimens, i.e. Mg-7SiC and pure magnesium to study the mechanism of wear. SEM micrograph of worn surfaces for Mg-7SiC composite and pure magnesium specimens at low load (5 N) has been presented in Fig. 4. Worn surfaces at higher load have been presented in Fig. 5.

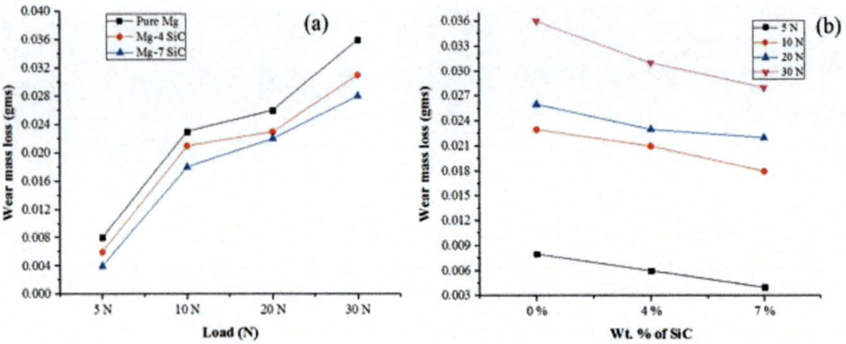


**Fig. 3** Mass loss of stir cast pure Mg and composites by varying **a** load, and **b** wt.% of SiC

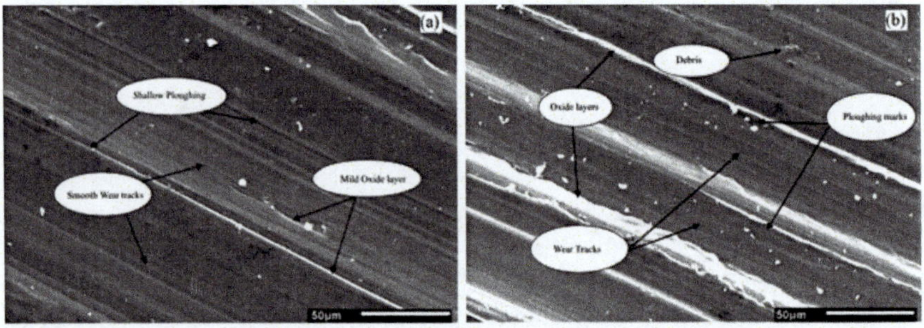


**Fig. 4** Worn surfaces of **a** Mg-7SiC and **b** pure Mg at 5 N load

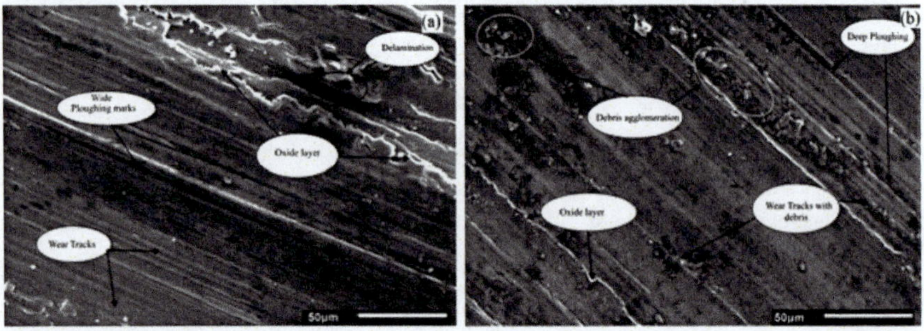


**Fig. 5** Worn surfaces of **a** Mg-7SiC and **b** pure Mg at 30 N load

Mg-7SiC composite worn surface at 5 N is presented in Fig. 4a. The wear surface presents smooth wear tracks, mild oxide layer and ploughing marks (shallow) at lower loads as can be seen in Fig. 4a. As load increased, wear tracks become rough and presence of oxidised surface increased for the Mg-7SiC composite as shown in Fig. 5a.

Oxidation wear is also considered a wear process in magnesium metal matrix. During wear test, specimens reached to higher temperatures, which induced the oxidation of the material [20]. Oxide layers are found throughout the sliding direction of wear. Delamination wear is considered as adhesive wear that occurs on the specimens owing to peripheral force which induce plastic distortion. As the plastic deformation crosses the acceptable limit, the crack originates and spreads to the boundary of Mg-SiC until the surface of the specimen is extracted [21]. The SEM images of pure Mg show oxide layers, ploughing marks at 5 N as presented in Fig. 4b. Figure 5b shows deep ploughing marks, debris agglomeration and wear tracks with debris in pure Mg at 30 N load. Transmitted materials, which have been corroded during wear test, have become the cause for debris formation. Debris may be the cause of abrasive wear that can affect the wear amount [22].

The oxide layers along the wear direction were found on all worn surfaces. Ploughing marks along the sliding path were caused by abrasion wear. Mg-7SiC composites have smoother wear track compared to pure Mg at 30 N load.

## 4 Conclusions

The following conclusions have been drawn from this study.

1. Magnesium-SiC composites produced by stir casting process shown the uniform distribution of particles.
2. The macrohardness of the Mg-7SiC composite was higher than the other specimens, as reinforcement avoids targeted deformation of the matrix material.
3. The wear resistance of the composites increased compared to pure magnesium under all load ranges. Mg-7SiC composite demonstrated higher wear resistance as compared to other specimens.
4. Ploughing marks and oxide surfaces were present in all worn specimens, although deep ploughing marks and debris agglomerations were present in worn surface of pure magnesium.

**Acknowledgements** The authors are thankful to IKGPTU, Kapurthala, for providing necessary help. The authors are also thankful to AICTE for providing the grant.

## References

1. Saravanan RA, Surappa MK (2000) Fabrication and characterisation of pure magnesium-30 vol.% $SiC_P$ particle composite. Mater Sci Eng A 276(1):108–116. https://doi.org/10.1016/s0921-5093(99)00498-0
2. Freeney TA, Mishra RS (2009) Effect of friction stir processing on microstructure and mechanical properties of a cast-magnesium–rare earth alloy. Metall Mater Trans A 41(1):73–84. https://doi.org/10.1007/s11661-009-0080-2

3. Ponappa K, Aravindan S, Rao PV (2013) Influence of $Y_2O_3$ particles on mechanical properties of magnesium and magnesium alloy (AZ91D). J Compos Mater 47(10):1231–1239. https://doi.org/10.1177/0021998312446501
4. Garcés G, Oñorbe E, Pérez P, Denks I, Adeva P (2009) Evolution of internal strain during plastic deformation in magnesium matrix composites. Mater Sci Eng A 523(1–2):21–26. https://doi.org/10.1016/j.msea.2009.06.026
5. Poddar P, Srivastava VC, De PK, Sahoo KL (2007) Processing and mechanical properties of SiC reinforced cast magnesium matrix composites by stir casting process. Mater Sci Eng A 460–461(Supplement C):357–364. https://doi.org/10.1016/j.msea.2007.01.052
6. Wang X, Wang N, Wang L, Hu X, Wu K, Wang Y, Huang Y (2014) Processing, microstructure and mechanical properties of micro-SiC particles reinforced magnesium matrix composites fabricated by stir casting assisted by ultrasonic treatment processing. Mater Design 57:638–645. https://doi.org/10.1016/j.matdes.2014.01.022
7. Bhiftime E, Guterres NF, Atmaja R (2017) Effect of wt.% SiCp and TiB on the mechanical properties in SiCp/AZ81A magnesium matrix composite by the method semi solid stir casting. Paper presented at the AIP conference proceedings. https://doi.org/10.1063/1.4968293
8. Aravindan S, Rao PV, Ponappa K (2015) Evaluation of physical and mechanical properties of AZ91D/SiC composites by two step stir casting process. J Magnes Alloys 3(1):52–62. https://doi.org/10.1016/j.jma.2014.12.008
9. Viswanath A, Dieringa H, Ajith Kumar KK, Pillai UTS, Pai BC (2015) Investigation on mechanical properties and creep behavior of stir cast AZ91-$SiC_p$ composites. J Magnes Alloys 3(1):16–22. https://doi.org/10.1016/j.jma.2015.01.001
10. Kandpal BC, Kumar J, Singh H (2017) Fabrication and characterisation of $Al_2O_3$/aluminium alloy 6061 composites fabricated by Stir casting. Mater Today Proc 4(2):2783–2792. https://doi.org/10.1016/j.matpr.2017.02.157
11. Rumiantseva YY, Bushlya V, Turkevich V (2019) The influence of SiC and $Al_2O_3$ whiskers on the properties of whisker-reinforced cBN-based composites. J Superhard Mater 41(6):377–387. https://doi.org/10.3103/S1063457619060029
12. Singh A, Bala N (2019) Relative sliding wear behavior of Mg metal matrix composites fabricated by stir cast route. Mater Res Express. https://doi.org/10.1088/2053-1591/ab4313
13. Sujayakumar P, Viswanath A, Kumar KKA, Rajan TPD, Pillai UTS, Pai BC (2013) Sliding wear behavior of stir cast AZ91/$SiC_p$ composites. J Solid Mech Mater Eng 7(2):169–175. https://doi.org/10.1299/jmmp.7.169
14. Gui M, Li P, Han J (2003) Fabrication and characterization of cast magnesium matrix composites by vacuum stir casting process. J Mater Eng Perform 12(2):128–134. https://doi.org/10.1361/105994903770343259
15. Ram B, Deepak D, Bala N (2019) Role of friction stir processing in improving wear behavior of Mg/SiC composites produced by stir casting route. Mater Res Express 6(2): https://doi.org/10.1088/2053-1591/aaf1e4
16. Muhammad WNAW, Sajuri Z, Mutoh Y, Miyashita Y (2011) Microstructure and mechanical properties of magnesium composites prepared by spark plasma sintering technology. J Alloy Compd 509(20):6021–6029. https://doi.org/10.1016/j.jallcom.2011.02.153
17. Chen L, Yao Y (2014) Processing, microstructures, and mechanical properties of magnesium matrix composites: a review. Acta Metallurgica Sinica (English Letters) 27(5):762–774. https://doi.org/10.1007/s40195-014-0161-0
18. Yu SR, Huang ZQ (2014) Dry sliding wear behavior of fly ash Cenosphere/AZ91D Mg alloy composites. J Mater Eng Perform 23(10):3480–3488. https://doi.org/10.1007/s11665-014-1157-7
19. Xiu K, Wang HY, Sui HL, Wang Y, Xu CL, Wang JG, Jiang QC (2006) The sliding wear behavior of TiCp/AZ91 magnesium matrix composites. J Mater Sci 41(21):7052–7058. https://doi.org/10.1007/s10853-006-0946-2

20. Wei TZ, Shamsuri SRB, Yee CS, Rashid MWA, Ahsan Q (2013) Effect of sliding velocity on wear behavior of magnesium composite reinforced with SiC and MWCNT. Proc Eng 68:703–709. https://doi.org/10.1016/j.proeng.2013.12.242
21. Huang S-J, Jeng Y-R, Semenov VI, Dai Y-Z (2011) Particle size effects of silicon carbide on wear behavior of SiCp-reinforced magnesium matrix composites. Tribol Lett 42(1):79–87. https://doi.org/10.1007/s11249-011-9751-4
22. Asadi P, Faraji G, Masoumi A, Besharati Givi MK (2011) Experimental investigation of magnesium-base nanocomposite produced by friction stir processing: effects of particle types and number of friction stir processing passes. Metallurg Mater Trans A 42(9):2820–2832. https://doi.org/10.1007/s11661-011-0698-8